BIS Publishers
Building Het Sieraad
Postjesweg 1
1057 DT Amsterdam
The Netherlands
T +31 (0)20 515 02 30
bis@bispublishers.com
www.bispublishers.com

ISBN 978 90 6369 479 1

Third printing 2019

Copyright © 2018 The University of Sydney School of Architecture and BIS Publishers.

All rights reserved. This publication may not be reproduced or transmitted in any form or by any means, electronic or mechanical, including photocopy, recording or any information storage and retrieval system, without permission in writing from the copyright owners. Every reasonable attempt has been made to identify owners of copyright. Any errors or omissions brought to the publisher's attention will be corrected in subsequent editions.

Martin Tomitsch
Cara Wrigley
Madeleine Borthwick
Naseem Ahmadpour
Jessica Frawley
A. Baki Kocaballi
Claudia Núñez-Pacheco
Karla Straker
Lian Loke

Design.
Think.
Make.
Break.
Repeat.

A Handbook of Methods

BIS

Table Of Contents

Foreword 7

Introduction 8

Methods 16
5 Whys 18
A/B Testing 20
Affinity Diagramming 22
Autobiographical Diaries 24
Bodystorming 26
Brainwriting 6-3-5 28
Business Model Canvas 30
Business Model Experimentation 32
Card Sorting 34
Cartographic Mapping 36
Channel Mapping 38
Co-design Workshops 40
Competitor Analysis 42
Contextual Observation 44
Cultural Probes 46
Decision Matrices 48
Design by Metaphor 50
Design Critique 52
Direct Experience Storyboards 54
Empathic Modelling 56
Experience Prototyping 58
Experience Sampling 60
Extreme Characters 62
Focus Groups 64
Forced Associations 66
Future Workshops 68
Group Passing 70
Hero Stories 72
Heuristic Evaluation 74
Interaction Relabelling 76
Interviews 78
KJ Brainstorming 80

◐ **Design.**
◐ **Think.**
◐ **Make.**
◐ **Break.**
◐ **Repeat.**

Laddering 82
Low-fidelity Prototyping 84
Mapping Space 86
Mindmapping (WWWWWH) 88
Mock-ups 90
Mood Boards 92
Online Ethnography 94
Perceptual Maps 96
Persona-based Walkthroughs 98
Personas 100
Questionnaires 102
Reframing 104
Research Visualisation 106
Role-playing 108
Scenarios 110
Science Fiction Prototyping 112
Service Blueprints 114
Sketching 116
Sketchnoting 118
Storyboarding 120
Thematic Analysis 122
Think-aloud Protocol 124
Usability Testing 126
User Journey Mapping 128
User Profiles 130
Value Proposition Canvas 132
Video Prototyping 134
Wireframing 136

Design Briefs 138
Autonomous Vehicles 140
Designing Space Travel 141
Museum Visitor Experience 142
Supermarket of the Future 143

Case Studies 144

Templates 162
5 Whys 164
Brainwriting 6-3-5 165
Business Model Canvas 166
Business Model Experimentation 167
Card Sorting 169
Channel Mapping 170
Competitor Analysis 171
Contextual Observation 172
Decision Matrices 173
Design by Metaphor 174
Extreme Characters 175
Focus Groups 176
Heuristic Evaluation 177
Online Ethnography 178
Perceptual Maps 179
Personas 180
Reframing 182
Role-playing 183
Scenarios 184
Science Fiction Prototyping 185
Sketching 186
Storyboards 187
Thematic Analysis 188
Value Proposition Canvas 189
Think-aloud Protocol 190
Usability Testing 191
User Journey Mapping 193
User Profiles 195
Wireframing 196

Design Team 196
Authors 198
Other Contributors 203

Glossary 204

Credits, Image Sources and References 206

Foreword

Design is a way of thinking that is relevant in our everyday life and our professional life. Everyday design thinking is a response to the needs and scenarios that arise in our personal day-to-day activities: deciding on the components of a meal, selecting the clothing we will wear, arranging the items on our shelves or in our rooms. As professional designers, we engage in a more complex process of design that draws on and reflects our education, previous experiences, and the application and adaptation of generalised methods. Formalised design methods date back more than 4,000 years to Babylonian times. Vitruvius, the Roman architect-engineer, 2,000 years ago, wrote a ten-volume treatise containing design methods for a wide range of specific problems in architecture and engineering of his day. Today we have moved from specific methods for individual problems to an array of methods that are applicable across a wide range of contexts. This book provides a collection of methods as a toolbox that can be adopted and adapted with ease by both professional and everyday designers.

Designers increasingly define the world we inhabit and affect our quality of life, our social life and our economic well being. Design is deeper than a style or an aesthetic that is applied after the engineering has been completed. Early approaches to design methods had a focus on the mapping from requirements and performance criteria to an efficient and cost-effective solution. With the design thinking movement, we are seeing design methods that include and go beyond thinking about the object or system being designed, to focus on the people that the design is intended to serve. This book, which is based on design research, practice and education in the Design Lab at the University of Sydney, includes a broad range of design methods that help the designer focus on how new designs can be responsive and mindful of the needs, desires, and mental models of people.

The Design Lab, and its precursor the Key Centre of Design Computing and Cognition, has its roots in the 1960s with a focus on the quality of design and the ways in which computational systems can improve our understanding of design as a creative process. Over the years, the faculty and students in the Design Lab have been influential in providing a strong theoretical and methodological basis for design cognition, AI models of design, computational creativity and, more recently, human-centred design.

This book represents another significant contribution to the community for both design education and practice. It provides a resource and a guide for novice and expert designers. As a guide for novice designers and designers still engaging in educational experiences, it provides a sense of scope and variety in the methods that can be applied at various stages in the design process. For expert designers, it is a resource that provides inspiration to try methods they may not have considered yet. The book represents a synthesis of methods described generally and in the context of case studies as well as a synthesis of views from authors with multiple perspectives on design education and theory.

John Gero and Mary Lou Maher
Founders of the Design Lab at the University of Sydney
October 2017

Design. Think. Make. Break. Repeat.

Introduction

Introduction

By Martin Tomitsch and Cara Wrigley

Design is no longer a discipline limited to the concerns of a singular, specific domain. Like most other industries, the field of design is being challenged by the arrival of the fourth industrial revolution. Systems are becoming more complex, requiring more intuitive user interfaces and multiple touchpoints, from wearable screens to virtual reality headsets. Digital systems are weaving their way into physical environments and products, from smart cities to the Internet of Things and medical devices. Technological advancements are changing the process of design. As a result, we must integrate the requirements of all the domains, aspects and features that make-up the most innovative solutions worldwide. We must design – think – make – break – and then repeat.

Design evolution

The responsibility of design has evolved over time along with industrial, technological and market shifts (Owen, 1991). For almost a century, design has been used to achieve a competitive advantage across industries. At the beginning of design as a profession, this involved designers working with engineers to achieve better construction techniques.

As markets changed and caught up with this trend, the role of design shifted to delivering a strategic advantage by having products with a better appearance, better human factors or usability, and better performance. Around the turn of the century, the role of design changed again, with companies seeking designers to help them develop better ideas and better integration, which now also included better experiences and social inclusion.

As our global and lived environments are becoming more complex, the role of design is changing yet again. We are facing unprecedented global challenges, such as population growth and mass-urbanisation, and technology is advancing and penetrating all aspects of our lives at a rapidly increasing rate.

Design is now seen as a pathway for solving complex, nonlinear problems, which can't be solved with technological or scientific approaches alone. It provides a framework for understanding the needs of the people, as well as the space to translate these needs into solutions. For the first time in the evolution of design as a field, the use of such methods is no longer limited to skilled design professionals. Using design as a way of thinking provides a strategic advantage across many professions. Design is, therefore, becoming a capability-enhancing skill, equipping people with the ability to deal with uncertainty, complexity and failure.

The last two decades have seen much excitement around the term 'design thinking', largely due to its adoption into business as an alternative approach to business strategy development. Herbert A. Simon first referred to design as a 'way of thinking' in his book *'The Sciences of the Artificial'* (Simon, 1969), proposing a structured approach for translating an existing situation into a preferred situation using design methods – helping to connect different elements contributing to a final solution. In the 1980s, the term 'design thinking' was used to describe the process of designing in architecture and urban planning (Rowe, 1991). Since then, several frameworks have been formulated to provide guidance for when, how and which methods to apply at the various stages of a design process. These early works laid the foundation for the role of design today as an innovation method.

Two popular design models that translate this way of thinking into a framework are the 'honeycomb' model proposed by the d.school at Stanford University and the 'double diamond' model published by the UK Design Council. The honeycomb model involves the stages of empathise, define, ideate, prototype and test, and stresses the importance of iteratively moving between those stages while working on a design project. The double diamond model entails the phases of discover and define (the first diamond), and develop and deliver (the second diamond). Each diamond encourages divergent thinking followed by convergent thinking. The first diamond focuses on understanding the problem, starting with a problem situation and ending with a problem definition. The second diamond uses the resulting problem definition as a design brief and is concerned with finding the right solution.

Despite being criticised by some scholars for their simplified view of design as a process, models like the honeycomb and double diamond offer distinctive perspectives and considerations. They allow organisations to adopt their own formalised design approach to inform how they operate and design their products and services.

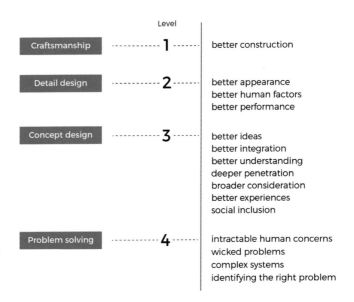

The changing role of design to provide a competitive advantage by achieving products, services, systems and environments of better quality. Levels 1 to 3 based on an original diagram by Owen (1990).

Design. Think. Make. Break. Repeat.

Design. Think. Make. Break. Repeat.

Arriving at an innovative solution is usually not a clear, straightforward pathway. Design requires learning about the context (the thinking part), building prototypes as tangible representations (the making part) and testing potential solutions (the breaking part). Rather than investing a lot of time in each step, it is more productive to go through the process as quickly and as often as possible (the repeating part). The earlier an idea or concept is broken, the quicker we can focus on improving it.

Design thinking

For an innovation to be successful, it is critical to have not only the technical and business opportunities in place, but also to ensure that there is a real need, a desire, for the product or service. According to Eric von Hippel, a Professor at the MIT Sloan School of Management, 70 to 80 percent of new product developments that fail do so not for lack of advanced technology, but rather a failure to understand users' needs. It's understanding who we are designing for and how to address their needs that companies find most challenging.

To understand who we are designing for (users, customers or other stakeholders), it is important to develop the skill of empathy. Design thinking uses a wide array of methods to develop empathy by collecting data from and about real people and translating this data into ideas and concepts.

Design making

Data and ideas collected during the design thinking phase can be turned into concepts and prototypes – the design making part of the process. This is where we build a tangible representation (or many representations) of the solution. In some cases, this is also referred to as the minimal viable product (MVP). A concept, prototype or MVP can be a representation of a specific scenario, the entire user interface, or just one feature built as a technical proof of concept.

The model for designing products or services used in this book. The methods are not limited to one of the phases; many of them can be applied at different stages of a design project.

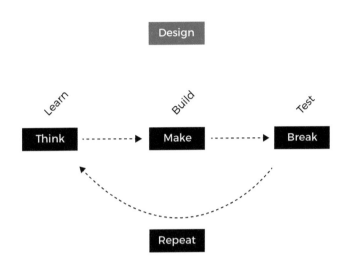

The steps of a design process are linked and interconnected. They don't happen in isolation from each other. The better the research data collected initially, the more useful the tangible representation of the solution will be.

Design breaking

One way to find out whether an idea works is to put it in front of potential users or customers. Sometimes it might be necessary to let go of an idea or concept to make room for even better ideas to emerge. To break a design solution requires embracing failure. Taking a different perspective and exploring many approaches rapidly can effectively solve complex problems.

In 1959, British industrialist Henry Kremer created a prize for designing a human-powered aircraft that could fly a figure eight course around two poles half a mile apart. Despite more than 50 official attempts, the prize went unclaimed for over 17 years. In 1976, Paul MacCready, an aeronautical engineer, completed the challenge by looking at the problem from a different perspective. While everyone else was trying to build a human-powered plane that can fly a figure eight around two poles, he built a plane that could be crashed and rebuilt within hours. His team would often break the plane several times a day, and from those failures learn how to improve their approach. The solution was to build a lightweight plane that could fly very slowly. Constantly breaking their concept sped up the process of finding a new, successful solution.

Repeating the steps

The final step is to repeat all or some of the previous steps. Every iteration leads to new insights, and the new insights are what will set a product or service apart from other solutions in the market. Designing, thinking, making and breaking many different representations quickly rather than striving to create one perfect solution leads to a more innovative outcome.

According to David Bayles and Ted Orland, a ceramics teacher one day announced that he would divide his class into two groups (Bayles & Orland, 2001). He explained to the groups that all those who sat on the left side of the studio would be graded based on the number of works they produced, while the right side would be graded based on quality. When it came to grading the students' submissions, he found that those focusing on quantity had come up with much more interesting and novel works than those striving to develop a high-quality submission. Not becoming fixated on one idea allowed the students to try out many different ideas quickly and so produce overall higher-quality works.

Who we are designing for

In interaction design, the end consumers of the designed products are commonly referred to as users. This notion is also reflected in the terms used to describe emerging design disciplines, such as user experience design, and methodologies like user-centred design. However, this is not always an accurate reflection of who is purchasing or engaging with an end design solution. Within the business and commercial world, the term customer is frequently used instead. In some cases, the user might not be the customer of a product. For example, users on Facebook are different from the customers, who are paying, for example, for targeted advertising. The design of Facebook as a platform needs to consider and target both. A design process may also need to consider other stakeholders, who are individuals or organisations with an invested interest or stand to gain or lose from the solution. The terms user, customer and stakeholders are not always equally interchangeable and have been carefully selected and used throughout this book.

How to use this book

This book is written as a learning resource and reference guide to scaffold the reader's understanding of the design process as a method for complex problem-solving and developing innovative solutions. The methods included in the book are applicable to a variety of design projects and across a range of domains and industries. This cross-perspective approach is also reflected in the choice of design briefs and case studies included in the book, which ranges from autonomous vehicles to designing the future shopping experience.

The book is divided into methods that include a full description along with step-by-step exercises and ready-to-use blank templates. The methods are included in alphabetical order, rather than structured by phases, to reflect that they can be flexibly used and adopted across multiple phases. Icons indicate the phases, in which each method is typically used. However, there is no hard rule about when a method can or cannot be applied.

Templates can be photocopied or used inside the book. The book is accompanied by a website (designthinkmakebreakrepeat.com), which provides printable versions of templates as well as further resources to illustrate the use of the methods.

As well as being a rich resource of design methods and materials, the book supports the teaching of students and readers from all disciplinary backgrounds. It provides everyday tools that assist with developing an understanding of design thinking by practically applying the methods through exercises. The methods included in the book have been contributed by leading experts in the field. The exercises are based on many years of experience in teaching the methods. All methods are grounded in research and include a list of academic articles that provide more detail.

The authors encourage researchers, practitioners and learners to use, modify, reinterpret and critique the contents of this book. We welcome any feedback, suggested improvements or experiences with successes – and most definitely failures! In the spirit of the book, we look forward to its ongoing development through conversations with you.

References

Bayles, D., & Orland, T. (2001). Art & fear: Observations on the perils (and rewards) of artmaking. Image Continuum Press.

Owen, C. L. (1990). Design education and research for the 21st century. Design Studies, 11(4), 202-206.

Owen, C. L. (1991). Design education in the information age. Design Issues, 7(2), 25-33.

Rowe, P. G. (1991). Design thinking. Cambridge, MA: MIT Press.

Simon, H. A. (1969). The sciences of the artificial. Cambridge, MA: MIT Press.

Design. Think. Make. Break. Repeat.

Methods

5 Whys

Uncovering root causes behind a problem statement

ACADEMIC RESOURCES:

Andersen, B., & Fagerhaug, T. (2006). Root cause analysis: simplified tools and techniques. Milwaukee, WI: ASQ Quality Press.

Collins, J. C., & Porras, J. I. (1996). Building your company's vision. Harvard business review, 74(5), 65.

Price, R. A., Wrigley, C., & Straker, K. (2015). Not just what they want, but why they want it: Traditional market research to deep customer insights. Qualitative Market Research: An International Journal, 18(2), 230-248.

Semler, R. (2004). The seven-day weekend: a better way to work in the 21st century. New York, NY: Random House.

The "5 whys" method helps to uncover a potential root cause to any surface level problem. The method provides a structured approach for repeatedly asking 'why' in order to provide deeper insight into the given problem. Originally developed by the Toyota Motor Corporation, this iterative method was first deployed to discover root cause analysis issues in manufacturing. Despite its engineering foundations, 5 whys is a popular design method identify any assumptions made and gain a deeper level understanding of a problem situation. The aim of continual questioning is to ensure that the right problem is examined and made central to the design process.

This method can be done independently by members of the design team or with stakeholders that have an involvement with the initial problem statement. For example, this can be a client or someone from a different team working on the same project. It is important that everyone involved in the method is familiar with the problem situation.

The method starts with a first-level surface problem statement, which should be based on findings from previous research activities. For example, an initial interview or questionnaire might have revealed a problem situation. Once the problem statement has been identified, we ask ourselves 'why' the problem occurs. In order to get to the root cause of the problem, we continue asking ourselves to explain the previous answer. Typically, the question 'why' is asked five times, but the number of iterations can be adapted until a satisfactory root cause is identified. The amount of questioning may also depend on the persistence of the person answering the question. The method is useful during the early phase of a design process to identify the right problem statement.

EXERCISE

YOU WILL NEED
Pen

In this exercise, you will practice the method of 5 whys and create a problem statement to summarise your findings. Use the provided template (p.164) to track the responses. Start at the top of the template and work your way down.

1 **Choose an initial problem statement** and write it into the 'challenge' section of the template.
E.g. most inner-city dwellers shop at the supermarket closest to their home. How can a supermarket brand become the preferred choice, rather than just the closest option?
[1 minute]

2 **Ask yourself why** this situation exists and **write down your response** in the first box. Answers should be longer than one word to provide enough detail to continue questioning.
E.g. people choose this supermarket because it is on the way home from work.
[3 minutes]

3 Ask **why for the second time**. Write down your response in the second box.
E.g. because they like the convenience of choosing what to buy each night.
[3 minutes]

4 Ask **why for the third time**. Write down your response in the third box.
E.g. because sometimes plans change at the last minute.
[3 minutes]

5 Ask **why for the fourth time**. Write down your response in the fourth box.
E.g. because changing plans at the last minute can lead to bought food going off in the fridge.
[3 minutes]

6 Ask **why for the fifth time**. Write down your response in the fifth box.
E.g. because wasting food is bad for the wallet and bad for the environment.
[3 minutes]

7 Once you feel you have **discovered the root cause of the problem**, describe it in detail and ideate some possible solutions to address it.
E.g. design a service for inner-city dwellers that takes away the importance of the supermarket's physical location by offering pick-up options suited to customers' needs and timetable.
[25 minutes]

Design. Think. Make. Break. Repeat.

A/B Testing

A is better than B ... or is it?

ACADEMIC RESOURCES:

Muylle, S., Moenaert, R., & Despontin, M. (2004, May). The Conceptualization and Empirical Validation of Web Site User Satisfaction. Information and Management, 41(5), 543-560.

Tullis, T., & Albert, B. (2013). Measuring the user experience: collecting, analyzing, and presenting usability metrics. Elsevier/Morgan Kaufmann (2nd ed., pp. 216-218).

A/B testing is an evaluation method that consists of testing in parallel two different versions of the same product, to decide which one is more suitable for a specific user need or business goal. It can also be used to assess a new version of an existing product against its predecessor. It is best suited to testing small, incremental changes in a design solution. By restricting changes to a known variable, it is then possible to understand the effect of that variable.

Even though this method is most commonly used to compare alternative versions of user interface designs for websites, A/B testing can also be applied to tangible products or prototypes. It is important to clarify the intended design objective behind the changes or features before testing. In outlining what you expect to happen, you have a point of measurement by which to test your hypothesis, for example, when A/B testing a simple versus a complex search function, you might predict that the simple one will be easier to use.

A/B testing is useful to offer a quick diagnosis on whether A or B works more effectively in a given scenario. For instance, when testing the location of a banner on a website, the metric of 'the time taken to complete a specific task' can reveal some comparable information in terms of visibility or accessibility. However, this information is not enough to reveal why one option is better than the other. The inclusion of post-experience interviews and other qualitative methods can offer insights beyond this metrics-based evaluation.

EXERCISE

YOU WILL NEED
2+ people, stopwatch, pen, paper

In this exercise, you will perform A/B testing using a pair of low-fidelity prototypes. The prototypes should only differ by a single variable, such as the size or location of a button. Focus on your own concepts, or use the resources on the companion website.

1. **Decide which variable to test** in your prototype. Write down a short statement summarising your hypothesis: what effect do you expect the variation to have?
E.g. a larger 'Submit' button is quicker to find.
[5 minutes]

2. **Prepare two sets of user interface sketches** for your prototype. The two sketches (A and B) should only differ regarding the design choice that you want to explore.
E.g. version A contains a large button. Version B contains a much smaller button.
[15 minutes]

3. **Choose the task that you want the user** to perform with your prototype.
E.g. complete and submit the flight bookng enquiry form.
[5 minutes]

4. **Select the evaluation metric** you will use to compare which version (A or B) is better. It should align with your hypothesis.
E.g. using time as a metric: How long does it take the user to complete and submit the form?
[5 minutes]

5. Give your first participant written instructions for the task. Ask them to **perform the task using version A first, then version B**. For each version, record the information that is relevant to your evaluation metric.
E.g. record the start and stop time of the task.
[20 minutes]

6. **Compare the results** from using version A and B by comparing your evaluation metric.
E.g. the task time = stop time - start time.

7. Repeat with your second participant, but **give them the tasks in the opposite order** (first version B and then version A). this counterbalances the effect that the testing order might have on the results. To make this method statistically significant, the test should be repeated with many users.
[20 minutes]

Design. Think. Make. Break. Repeat.

Affinity Diagramming

Translating research data into user needs

ACADEMIC RESOURCES:

Holtzblatt, K., Wendell, J.B., & Wood, S. (2005). Chapter 8, Building an Affinity Diagram. In Rapid Contextual Design, Burlington, MA: Morgan Kaufmann.

Thorough user research doesn't end with the collection of data. It is equally important to analyse the data in order to derive insights about the design problem. User research methods such as interviews (p.178) can lead to large amounts of data, which makes it difficult to gain insights just by looking through the data, for example, listening to audio recordings or reading through transcripts.

Affinity diagramming is a simple and cost-effective systematic method for processing such data, which is typically qualitative in nature. It allows for analysing data (breaking down of data into parts) as well as synthesis (forming a coherent whole out of the parts).

Ideally, an affinity diagram is generated by a group of people, consisting of designers as well as other stakeholders. Participants collectively go through the data and identify specific problems or observations, which are written down on a yellow Post-it note, called an affinity note. The aim is to create as many notes as possible by only recording one aspect per Post-it. These notes are then grouped on a wall according to common themes and labelled through a blue Post-it note per cluster. The blue notes are written in first person from the perspective of the user. Depending on the amount of data, this step is repeated to cluster the blue notes, which are then abelled with pink notes, again written in the first person and further abstracting the data.

The final step is to 'walk the wall' in order to generate ideas. In this step, concrete ideas for potential solutions are recorded on green Post-it notes and attached to specific pink, blue or yellow Post-it notes.

EXERCISE

YOU WILL NEED
2–6 people, pen, paper, Post-its (yellow, blue, pink, green), wall

In this exercise, you will analyse data from an interview transcript using a 'bottom-up' approach. Use the affinity diagramming method to cluster the data so that themes and patterns emerge. Focus on your own design problem, or use the resources on the companion website.

1 Each person **reads a different section of the interview data**. Highlight the text as you read, looking for statements in which users express their interests, needs, issues and motivations.
[30 minutes]

2 **Make affinity notes on yellow Post-its**. Each note should record an observation found in the data. Write a single observation on each note. Aim to create lots of notes, and place them on the wall when you are done.
[10 minutes]

3 As a group **cluster the yellow Post-its** by apparent similarity. Rearrange them so that similar observations are in a single column. There should be 3 to 6 notes per column; if there are more you might be burying a distinction that needs to be split into a new column. Aim to create lots of clusters.
[10 minutes]

4 **Use a blue Post-it note to label each cluster**, placing it at the top of the column. Blue notes should be written as a need expressed in the voice of a user.
E.g. 'I want to find things quickly without having to search for them.'
[10 minutes]

5 **Start adding pink Post-its.** Move blue labels together that seem to have related themes. Place a pink Post-it above the group and write a unique label on it. The language of the pink labels is also in the voice of the user.
[10 minutes]

6 **Review the affinity diagram** by walking along the wall and reading it from the top down. Change labels and shift notes around if needed, to create a more powerful set of user needs and insights.
[10 minutes]

7 **Identify potential solutions** addressing specific Post-it notes and add them using green Post-its.
[5 minutes]

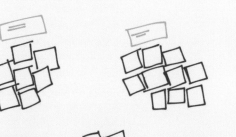

Design. Think. Make. Break. Repeat.   23

Autobiographical Diaries

A starting point to think about the lives of others

ACADEMIC RESOURCES:

Breakwell, G.M. (2006). Using Self-recording: Diary and Narrative Methods. Research Methods in Psychology (pp. 254-273). Sage.

Carter, S., & Mankoff, J. (2005, April). When participants do the capturing: the role of media in diary studies. In Proceedings of the SIGCHI conference on Human factors in computing systems (pp. 899-908). ACM.

Go, K. (2007). A scenario-based design method with photo diaries and photo essays. Human-Computer Interaction. Interaction Design and Usability, 12th International Conference, Proceedings, Part I (pp. 88-97). Springer.

Neustaedter, C., & Sengers, P. (2012, June). Autobiographical design in HCI research: designing and learning through use-it-yourself. In Proceedings of the Designing Interactive Systems Conference (pp. 514-523). ACM.

Diaries are a well-established method of recording self-reported data from users. Compared to other methods, such as questionnaires (p.102) or interviews (p.78), the usage of diaries has the advantage that users record events when they occur. In design research, diaries are typically used to understand how people complete daily activities, in order to direct or evaluate the design of new products or services.

The use of autobiographical diaries has a long history. People have documented their personal experiences in journals and notebooks, and, more recently, using digital media such as blogs. The process of self-documentation can assist with developing a deeper understanding of one's own practices. During a design process, self-documentation represents a valuable tool to reflect on how others might experience a product or service.

Autobiographical diaries, when used in design, document the use of everyday products or services through textual and visual means. The use of autobiographical diaries is intended to offer an additional perspective, or a starting point, rather than to function as a stand-alone method for data collection. As such, autobiographical diaries complement research methods geared towards forming an objective view of users' needs, which removes the designer's personal opinion from the equation. However, informal practices such as self-testing do occur implicitly as part of every design process – typically before designs are tested by others. Documenting this process of self-experience can, in fact, be an opportunity for further reflection.

Autobiographical diaries can be used during the initial phases of the design process, either to gain inspiration or to find general issues that can be explored further through other design methods.

EXERCISE

YOU WILL NEED
Pen, paper, camera, word processing software.

In this exercise, you will select a piece of household technology to reflect upon. Alternatively, you may choose an object relevant to your own design problem..

Document your reflections twice a day for five days.

1 **Write a description** at the start of each diary entry including the following pieces of information:
- Date and place: *E.g. when and where the situation takes place*
- Description of the context: *E.g. the environment of use*
- Aesthetics of the experience:
- *E.g. feelings, memories triggered, associations*
- Reflective accounts: *E.g. Which features of the object are most memorable and why? How does the object allow you to perform your tasks?*
- Subjectivity: *E.g. how might your biases as a designer and as a user differ?*

[30 minutes]

2 **Form your own critical questions** in response to the situations that emerge from your experience with the object. Write them down, and then reflect upon and respond to these questions in your diary entries.

3 **Describe your experience using images**. Think of photographs as tools that help you tell a story and **take photos of the processes and objects** that form a part of your experience. Write small captions and make annotations on the images to help you describe those narratives.

4 After **finalising your diary** (with at least ten reflections over five days), take some time to **conclude and reflect** on the lessons learned. Write these reflections down.
Some examples of relevant questions to ask yourself might be:
- Have you learned something new about the object you have been interacting with?
- How do you think that becoming aware of your behavioural patterns influenced the way you interacted with this particular object throughout the process?
- Did you have any realisations during the documentation process?

[20 minutes / 5 days for recording reflections]

Design. Think. Make. Break. Repeat.

Bodystorming
Thinking with your body

ACADEMIC RESOURCES:

Burns, C., Dishman, E., Verplank, W., & Lassiter, B. (1994, April). Actors, hairdos & videotape – informance design. In Conference companion on Human factors in computing systems (pp. 119-120). ACM.

Oulasvirta, A., Kurvinen, E., & Kankainen, T. (2003). Understanding contexts by being there: case studies in bodystorming. Personal and ubiquitous computing, 7(2), 125-134.

Schleicher, D., Jones, P., & Kachur, O. (2010). Bodystorming as embodied designing. Interactions, 17(6), 47-51.

"Bodystorming" is a form of brainstorming, with an emphasis on generating ideas and unexpected insights through physical exploration, experience and interaction. While brainstorming is typically done with pen and paper sitting down, bodystorming requires active physical participation. Using your whole body to act out situations enables you to draw inspiration from the way in which your senses respond to the world, not just how you think about it.

This tacit knowledge is not always readily available when talking about how we experience the world, but can be accessed through acting out situations collaboratively. These shared experiences can help to improve our understanding of how people experience the products or services that we design.

Bodystorming can be used to explore and understand existing practices, leading to the identification of current issues and opportunities for design ideas. Starting with familiar situations can help participants to develop the necessary mindset for envisioning future scenarios and practices.

Techniques from theatre and dramatic improvisation can be borrowed to help explore existing and alternative situations. What-if scenarios, props and role-playing can help designers and users to collaboratively explore and simulate the use and experience of real and envisioned products or services. It is important to physically warm-up before doing bodystorming, as it requires thinking generatively through physically acting out experiences. An important aspect of this method is therefore to enjoy being playful.

EXERCISE

YOU WILL NEED
3–4 people, stopwatch, marker, paper, masking tape, scissors, furniture, camera, Post-its

In this exercise, you will explore an existing situation through physically acting out scenarios, with the goal of revealing unexpected insights and discovering opportunities for design. Use the resources on the companion website to help you keep the bodystorm going.

1. **Brainstorm about the problem situation:** waiting for a doctor's appointment. Think of at least three different scenarios that could take place. Note down any ideas that come up, which you can explore during the bodystorm. Record issues and ideas on one sheet of paper and scenarios on another.
   [10 minutes]

2. **Set up the physical space** to simulate a doctor's waiting room, using the furniture, masking tape and paper mock-ups (signs, screens, objects, etc.).
   [10 minutes]

3. **Assign roles** to each person: an actor or an observer:
   - Actor: Two or more people interpret the scenario and role-play the characters. Use Post-it notes to label actors. Actors can also play the role of things, such as a reminder device. The actors improvise to generate the action.
   - Observer: One person observes, takes notes, makes sketches, and documents the bodystorm with photos. Observers can call out triggers from the cue sheet.

4. **Act out** a concrete scenario: the waiting room is full when you enter. You have a fever. You are informed there will be an hour-long wait …
   [15–20 minutes]

5. **Use props to generate solution ideas** for problems that emerge, for example, imaginary devices. You can start with a simple sketch on a piece of paper. As the scenario evolves and user needs are identified, the imaginary device can be modified by altering its form, function and behaviour. Quickly sketch the changes - and incorporate them into the scenario enactment.

6. As the role-play continues, **try the 'Freeze' and 'What-if' triggers.**

7. **Switch roles** and repeat for the different scenarios.
   [15–20 minutes]

8. **Discuss and document:**
   - What did you learn about the problem situation? Did you discover anything new, different or unexpected from bodystorming compared to brainstorming?
   - What opportunities for design did you discover through bodystorming?
   [10 minutes]

Brainwriting 6-3-5

Building on each other's ideas

ACADEMIC RESOURCES:

Rohrbach, B. (1969). Kreativ nach Regeln – Methode 635, eine neue Technik zum Lösen von Problemen [Creative by rules - Method 635, a new technique for solving problems]. Absatzwirtschaft (Vol. 19, pp. 73-75).

"Brainwriting 6-3-5" is a form of brainstorming that was developed as a way of getting around the group dynamic issues that can plague conventional brainstorms (Rohrbach, 1969). For example, shy people do not contribute as much as they have to offer, strong personalities dominate the conversation and existing power relationships (e.g. between employees and managers) can influence the ideation process. Brainwriting 6-3-5 overcomes these issues by combining individual and collaborative ideation. Like brainstorming, it aims to stimulate creativity in the early phases of a design process.

Brainwriting 6-3-5 received its name from the set-up of the session: six team members record three ideas in five-minute cycles. In the first round, each participant records their idea in the first row on a piece of paper. At the end of the five-minute cycle, they pass it to the team member on their left. In the second round, each team member reads the entry from the previous round and records three new ideas in the second row that each build on the idea from the row above. This step is repeated until the sheet returns to the person that started the first row. That way people are encouraged to build on each other's ideas by reviewing what has already been recorded and adding to it or changing it. Brainwriting 6-3-5 can yield more ideas in less time than a conventional brainstorm session, up to 108 ideas in 30 minutes.

The brainwriting 6-3-5 session should be started with a general discussion of the problem area between the participants (potentially guided by a moderator). This ensures that all team members are aligned regarding the topic to be tackled in the session.

EXERCISE

YOU WILL NEED
4–6 people, pens, paper

In this exercise, you will generate a range of concepts by building on each other's ideas. Use the provided template (p.165) or an A4 sheet folded into 3x6 rectangles. You can also vary the number of people, e.g. brainwriting 4-3-5 with four people.

1. **Choose a topic** for your brainwriting session: a design problem that you would like to solve. If you don't have a topic at hand, you can choose one from the design briefs (p.138). Discuss the topic with your group.
[5 minutes]

2. **Record three different ideas** in the top row of the brainwriting sheet. The ideas should be possible solutions related to your design topic, and draw on your knowledge of user needs within this area.
[5 minutes]

3. **Pass your brainwriting sheet** to the person on your left, then begin the next round.

4. **Review the sheet** you have received and the ideas recorded by the person before you. In the next row, record three more solutions inspired by what the person before you has written. Try out the following options:
   - Recording new ideas
   - Adapting the existing ideas
   - Combining ideas with each other
   - Modifying or adding to ideas.
[5 minutes]

5. **Repeat the process** until everyone has recorded three ideas on each of the brainwriting sheets and the sheets have returned to their original owners.
[20 minutes]

6. **Present some of the ideas**. Each person picks their favourite idea recorded in their own brainwriting sheet and explains it to the group. You can also cut the sheet so that each idea is represented on an individual piece of paper, which allows for collaboratively sorting the ideas.
[10 minutes]

Design. Think. Make. Break. Repeat.

Business Model Canvas

Visually designing the value a company offers

ACADEMIC RESOURCES:

Amit, R., & Zott, C. (2012). Creating value through business model innovation. MIT Sloan Management Review, 53(3), 41.

Osterwalder, A., & Pigneur, Y. (2010). Business model generation: a handbook for visionaries, game changers, and challengers. John Wiley & Sons.

Wrigley, C., Bucolo, S., & Straker, K. (2016). Designing new business models: blue sky thinking and testing. Journal of Business Strategy, 37(5), 22-31.

Several studies show that business model changes are among the most sustainable forms of innovation. For example, a company like Uber would not have been able to disrupt the taxi industry just by having a great user interface for their mobile app. Although their mobile app was an innovative product, it is their business model that led to their widespread popularity. When designing new products or services, it is therefore important to also consider the design of the underlying business model. Just like products and services need to be iteratively developed and tested, good business models also need to go through the stages of conceptual development, exploration and implementation.

The business model canvas provides a template for visualising the value a company offers to their customers. Used in the stage of conceptual development, it can assist in identifying important opportunities for innovation not usually addressed by product or process development alone. The business model canvas includes nine building blocks (see the provided template on p.166), which reflect the constituent elements of a company's business model and offer insights into the model's metrics and aspects that can be manipulated.

The method allows rapid mapping of the elements of a company and the value it offers, before a product or service is thought about. In the conceptual stage of idea generation, it can allow for multiple options to be presented and evaluated in a short period of time. In order to hold a strong competitive position in the market, it is the combination of these designed elements (the nine building blocks) that will make it hard for competitors to copy a product or service offering.

EXERCISE

YOU WILL NEED
Pen, internet access

In this exercise, you will fill out the business model canvas for an existing company using the provided template (p.166). If you don't have a specific company in mind, select one related to a design brief of your choice (p.138).

1. **Find your chosen company's** mission statement and write this as the value proposition. What product or service do they offer the users?
[5 minutes]

2. **What kind of customer is the company targeting?** Place this information in the customer segments block. Do they serve a specific niche or the mass market? Do they target several segments simultaneously by acting as a middleman?
[5 minutes]

3. **What relationship does the company have with their customers?** Place this information in the customer relationship block. Do they give their customers a framework to carry out self-service, or offer dedicated personal assistance?
[5 minutes]

4. **How does the company reach its customers?** This could be a real or digital location, like a store or a website. The channel might vary over time; depending on whether the customer is browsing, purchasing, or using their new product.
[5 minutes]

5. **How does the company make revenue?** There are many ways to do this beyond simply selling a product, such as renting, licensing or offering subscription fees.
[5 minutes]

6. **What key activities does the company need to perform?** These are ongoing tasks, such as producing products or maintaining a platform for their customers.
[5 minutes]

7. **What resources are required to run the company?** For example, physical equipment, intellectual capacity, human resources or financial resources.
[5 minutes]

8. **Who are your partners?** A company doesn't need to do everything on its own. They may create strategic alliances, outsource work, or build strong buyer-supplier relationships.
[5 minutes]

9. **What does it cost?** What are the most important costs in the business model design? Are they fixed costs, or do they vary over time?
[5 minutes]

10. **Review your finished business model.** Where are the weaknesses? Which parts work well? Create a second iteration where you tackle the weaknesses and emphasise the strengths.

Design. Think. Make. Break. Repeat.

Business Model Experimentation

Iteratively exploring ideas for business model designs

ACADEMIC RESOURCES:

Teece, D. J. (2010). Business models, business strategy and innovation. Long range planning, 43(2), 172-194.

Wrigley, C., & Straker, K. (2016). Designing innovative business models with a framework that promotes experimentation. Strategy & Leadership, 44(1), 11-19.

Sosna, M., Trevinyo-Rodríguez, R. N., & Velamuri, S. R. (2010). Business model innovation through trial-and-error learning: The Naturhouse case. Long range planning, 43(2), 383-407.

When designing products or services, we try to come up with many ideas and explore the envisioned product or service from many perspectives. The same approach can be brought to designing business models, which are the backbone of any company that offers products or services. Business models need to be customer-centred, as they describe what value the products or services bring to the customer of a company. However, the customers are not always the users of those products or services. For example, in a social networking site, such as Facebook, the users enjoy the service and the customers generate value for the company by paying for advertising. The business model needs to be designed to consider both.

It is easy but dangerous, for a company to decide on a business model too quickly without iterating their approach first.

Business models need to be well-designed to ensure they consider all aspects of a company and their products or services. Building upon the business model canvas method (p.30), business model experimentation provides a structured exploration approach to testing new ideas by exploring different aspects. It presents five distinct areas of the business model canvas to focus on and experiment with. Changing the focus area can inspire new ideas about how to provide value to the customers, which, in turn, can lead to the creation of alternative business models. Studies found that a trial and error process in the early stages of business model innovation is critical to the success of a company (Sosna et al., 2010). By controlling different parts of the business model canvas, the business model experimentation method allows for contrasting different scenarios before rolling out a new business plan.

EXERCISE

YOU WILL NEED
Pen, internet access

In this exercise, you will generate diverse business model concepts based on five different foci. Use the provided templates (pp.167–168) and the resources on the companion website to guide you. Begin with a business model canvas (p.30) filled in for a company of your choice, or use the resources provided on the companion website.

1. **Fill in the customer-led template** (p.167) by tackling each of the nine building blocks of the business model canvas in the order shown on the template. This allows you to leverage the design of the remaining blocks for one known aspect: your current customers. A customer-led focus explores the diverse possibilities that lie within new and untouched customer segments.
[10 minutes]

2. **Fill in the cost-driven template** (p.168) in the order shown on the template. A cost-driven focus looks at ways of reducing expenses in order to find opportunities elsewhere.
[10 minutes]

3. **Compare and evaluate** the two resulting business models. What are their main differences? What are the strengths of each one? Take these reflections with you into the next step of the exercise.
[10 minutes]

4. **Rework the canvas multiple times**, starting with a different focus each time. The templates for resource-led, partnership-led and price-led business models are available on the companion website.
E.g. a resource-led focus identifies ways in which a business can gain better value from its existing resources, by exploring novel ways to restructure or reapply them.
E.g. a partnership-led focus is the exploration of new resources and capabilities from external partnerships.
E.g. a price-led focus reduces the cost across the entire business model, in order to provide the same value at a much cheaper price.
[30 minutes]

5. **Compare and evaluate the resulting concepts**, based on the five completed templates. If further exploration is required, repeat this process.
[15 minutes]

Design. Think. Make. Break. Repeat.

Card Sorting

Seeing information from your user's perspective

ACADEMIC RESOURCES:

Courage, C., & Baxter, K. (2006). Understanding your users: a practical guide to user requirements methods, tools, and techniques. San Francisco: Morgan Kaufmann Publishers.

Sinha, R., & Boutelle, J. (2004, August). Rapid information architecture prototyping. In Proceedings of the 5th conference on Designing interactive systems: processes, practices, methods, and techniques (pp. 349-352). ACM.

From newspapers to websites, microwave ovens to mobile apps, the average person encounters and navigates a huge amount of information in the form of text, icons, images and numbers on a daily basis. Whether in the interface of a simple alarm clock or the workings of a large complex multi-user system, the way information is labelled, grouped and structured shapes how a product or service is used. Designers often refer to this structuring of content as a product's information architecture. This is most clearly seen in website design, where navigation, tabs and pages create hierarchies and groups that order what the user sees.

Card sorting is a method that allows users and stakeholders to participate in the design of an information architecture. As its name suggests, card sorting involves participants sorting and organising cards into groups that are meaningful to them. In giving participants tangible cards, this method can provoke discussion and action that can help:

• Discover what information or tasks should be included or excluded in a design
• Discover what terminology users know this by (e.g. 'News' or 'Current Affairs')
• Design new ways for this grouping and structuring of informatin in a product or service.

Card sorting can help us better understand our user's mental model and build information structures that 'talk' to this. Just like a building, good information architecture can help users know where they are and where they are going. As card sorting requires participants to sort existing topics, it is a method best suited to refining a new (but existing) concept or redesigning an existing product.

EXERCISE

YOU WILL NEED
4 people, scissors, pen

In this exercise, you will carry out card sorting to inform the redesign of an existing website or mobile app. Use the template on the companion website to create your cards and the provided template (p.169) to take notes. Use an existing website or app, or choose a site map from the resources on the companion website.

1. **Form two groups of two.** Each group will design and facilitate a card sorting activity for the other group, so that you can experience card sorting from the position of both facilitator/designer and participant.
[2 minutes]

2. **Choose an appropriate website or app** as the basis for your card sorting activity. Make sure it is something the other group is likely to use.
E.g. website – group one: Flight booking portal (from the companion website)
E.g. website – group two: Online grocery shopping (from the companion website)
[4 minutes]

3. **Choose a user goal and sub-section of the site or app** to create cards for. You should focus on a clear task
E.g. buy a loaf of bread from the bakery section
[4 minutes]

4. **Create your cards** by writing all the existing names and categories the chosen sub-section of the website uses. These words might be found in the navigation (tabs, pages, headings) or in search filters and tools on your chosen website/app.
E.g. online groceries: bakery, deli, freezer, drinks, liquor, tobacco, etc.
[5 minutes]

5. Place all cards on the table (including blank ones) in a random order. **Invite your participants (the other group) to sort the cards into groups and hierarchies.** As the facilitator:
   - Brief the participants on the product and user goal
   - E.g. buying a load of bread
   - Invite participants to 'think aloud' (p.124)
   - Record the results of the card sort activity using the template (p.169)
[15 minutes]

6. **Translate your findings into recommendations.** For your website or app, what would you change? What would you keep the same? Why?
[10 minutes]

7. **Swap roles and repeat.** What did you learn?

Design. Think. Make. Break. Repeat.

Cartographic Mapping

Generating rich depictions of settings and practices in a problem domain

ACADEMIC RESOURCES:

Elovaara, P., & Mörtberg, C. (2010). Cartographic mappings: participative methods. In Proceedings of the 11th Biennial Participatory Design Conference (pp. 171-174). ACM.

Finken, S., & Mörtberg, C. (2014). Performing Elderliness-Intra-actions with Digital Domestic Care Technologies. In IFIP International Conference on Human Choice and Computers (pp. 307-319). Springer, Berlin, Heidelberg.

Mapping, and other methods involving making collages, are frequently employed in participatory design workshops to capture and understand domain specific user knowledge. Cartographic mapping is a mapping method with a particular focus on the mediating role of the mapmaking activity in mutual knowledge construction. In this method, the facilitator and participant work together to create a visualisation of the participant's daily routines, relationships and settings within a problem domain.

A typical cartographic mapping process involves two stages taking place in a workshop setting: making an initial map, and enhancing the map through a participant-performed ethnographic study. In the first stage, workshop participants are asked to create a map of their relationships with other people, devices, and other material objects in their problem domain. A large blank piece of paper, various cut-out pictures, Post-it notes and coloured markers are provided for the activity. The participants place a picture representing themselves on the paper, and then start to map relations with other entities around it. During this process, the workshop facilitator asks questions about the participants' particular choices of images and the relationships being mapped. In the second stage, the participants are asked to take photographs of the setting relating to the problem domain to capture the details of their work or everyday routines. In a subsequent workshop, the participants add these photographs to the maps they created. to develop a better understanding of the problem domain.

In addition to the creation of rich visual representations of people's daily routines, relationships and settings, the activity of mapmaking facilitates an informal conversation, supported by relevant visuals, about the various problems and matters of concern.

EXERCISE

YOU WILL NEED
1–3 people, A0 and A4 paper, coloured markers, pen, scissors, glue, sticky tape

In this exercise, you will employ cartographic mapping to understand the practices of one or more participants and identify opportunities for design solutions. Focus on your own design problem, or follow the 'Supermarket of the Future' brief (p.143) and use the resources on the companion website.

1. Arrange a workshop with one or more participants. Every participant should have recent experience of the problem domain.
E.g. shopping for groceries in the supermarket

2. Ask the participant(s) to use the provided images and materials on the companion website to express their experiences. They can glue these onto the A0 paper, arranged in such a way that they represent the participant's routines and relationships. If you are focusing on your own design problem, you will need to create these resources yourself, using the example on the companion website for inspiration.
[15 minutes]

3. Your participant(s) can also use lines, annotations and sketches to accompany the pictures they have selected.
E.g. A line connecting two pictures could represent a relationship
E.g. Annotations can be used to clarify the choice of a picture.
[10 minutes]

4. Use the resulting map to interview your participant(s). Ask them questions about their activities, the people they interact with, the technologies they use, and the problems they face. Follow up any interesting points that you observed during the map-making. Take notes and/or record the conversation.

5. During the week after the workshop, **ask your participant(s) to take photographs** of the environment, objects and technologies they encounter in the problem domain. Print these photos.
[1 week]

6. Conduct a second workshop where you ask the same participant(s) to augment their existing map with the photos they took. This will help to improve the representation and understanding of the problem domain.
[20 minutes]

Design. Think. Make. Break. Repeat.

Channel Mapping

Reaching your customers from all angles

ACADEMIC RESOURCES:

Straker, K., Garrett, A., Dunn, M., & Wrigley, C. (2014). Designing channels for brand value: four meta-models. In Bohemia, Erik, Rieple, Alison, Liedtka, Jeanne, & Cooper, Rachael (Eds.) Proceedings of 19th DMI: Academic Design Management Conference (pp. 411-431). London: London College of Fashion.

Channel mapping is a method that explores what a company's brand values mean to a particular customer. It does this by going beyond listing the ways in which customers are contacted by that company. Companies engage with their customers through a range of channels. Every interaction that people have with a company involves some form of channel. Channels can be digital (e.g. websites) or physical (e.g. stores). They represent routes of communication between an organisation and its customers, which includes completing commercial transactions. In particular, online purchases are one of the most rapidly growing forms of shopping, with the growth rate of online sales outperforming buying through traditional retail channels.

As channels diversify, customers engage with an increasing number of channels at once. It is, therefore, necessary to carefully design how people experience each individual channel as well as how this relates to the overall experience and interactions between people and a company. It is critical to understand how customers behave and their motives in order to design a relevant customer experience across all channels. Using more than one channel means thinking strategically about how each channel can help shape an individual's experience with a particular company or brand.

The channel mapping method assists with the creation of the overall customer experience. It involves several elements, such as brand value and meaning, which are explored to determine how customers experience or interact with them. Through designing and assessing each element in the tool, it is possible to quickly explore and evaluate alternative channel designs that not only reach the customers but also align with the brand value communicated.

EXERCISE

YOU WILL NEED
Pen, paper, internet

In this exercise, you will map out the channels of a company of your choice, using the provided template (p.170). The company could be related to a design project you are currently working on or one that you are familiar with from your personal experience.

1 Choose a company and **come up with a list of their brand values.** A brand value refers to a particular attribute of that company, expressed in a way that a customer can easily identify with.
E.g. TripAdvisor = convenience, freedom
E.g. Burberry = high-quality, luxury
E.g. Whole Foods Markets = community, integrity
[5 minutes]

2 **Choose one type of customer.** The customer could be from the customer segment currently targeted by that company, or part of a potential segment they could reach. Think about what kind of people typically purchase this brand.
E.g. elderly, millennials, brand buyers, bargain hunters
[10 minutes]

3 **Describe the meaning of your chosen brand value to your chosen customer.** For example, the brand experience of 'community' through the lens of an elderly customer could translate to 'support'.
[10 minutes]

4 I**dentify all the key channels** that could potentially be used to reach that customer. Common channels include websites, catalogues, physical stores, kiosks, social media and local events.
[15 minutes]

5 **Think about how channel can be tailored** to reflect the chosen brand value For example, a grocery website that reflects the value of 'support' might offer diverse delivery options, a support function and very clear step-by-step instructions.

6 Once you have finished mapping out one channel, the process can be repeated, **mapping out a different channel experience** for the same customer and brand value.

Design. Think. Make. Break. Repeat.

Co-design Workshops

Designing with your participants

ACADEMIC RESOURCES:

Sanders, E. B. N. (2002). From user-centered to participatory design approaches. Design and the social sciences: Making connections, 1(8).

Steen, M., Manschot, M. A. J., & De Koning, N. (2011). Benefits of co-design in service design projects. International Journal of Design 5 (2) 2011, 53-60.

Co-design workshops bring users, customers, stakeholders and designers together to rapidly and iteratively critique design concepts, ensuring that the needs of the people we are designing for remain at the centre of the design process. Co-design and similar methodologies, such as participatory design, involve the users and other stakeholders participating actively, building on concepts they are presented with (be it a current user experience or a new design concept) and informing the future direction of the design. The principle of co-design is to 'design with' rather than 'designing for' people. Users and other stakeholders are in an active role, contributing to the design, rather than passively responding to design decisions.

Co-design workshops build on this principle and include a preparation phase, recruitment phase, the workshop itself, interpretation and action. The first phase, preparation, is used to determine the overall direction for the workshop.

This can involve the development of an initial concept that users can respond to, for example, in the form of a low-fidelity prototype (p.84) or a storyboard (p.120). During the workshop, participants are taken through the stages of immersion, talking about current experiences, ideal experiences, and finally, evaluating and iterating the initial concept. Comments from participants along with any artefacts that were co-designed during the workshop are then analysed and fed back into the design process.

Co-design workshops can be employed at any stage of the design process. During the research phase, they can be used to inform a complete view of people's circumstances and situations. For projects that focus on the redesign of an existing product or service, this includes developing an understanding of how people currently make use of the product or service. During the prototyping phase, co-design workshops can be used to rapidly iterate concepts.

EXERCISE

YOU WILL NEED
3+ people, pens, paper

In this exercise, you will learn how to design and conduct a co-design workshop. You will decide what the purpose of the workshop is, who the participants are and which methods to use. Focus on your own design problem, or follow the 'Supermarket of the Future' brief (p.143).

1 **Decide what you want to achieve** from the co-design workshop and write it down.
E.g. a better way of buying fresh produce
[5 minutes]

2 **Consider the logistics** of your co-design workshop:
- What kind of people should be there?
  E.g. frequent shoppers, avid cooks
- How will you record it?
  E.g. notes, written feedback, observations, video
- What is the order of activities and their duration?
  E.g. immersion, talking about current experiences, describing ideal experiences, evaluating initial concepts.
- Prepare a script.

[20 minutes]

3 **Prepare the workshop materials**. Use printed images to immerse participants in the problem space. Leverage existing sketches, or prototypes of initial concepts, or select examples from the resources on the companion website for 'Supermarket of the Future'. Identify methods to complete during the workshop such as:
- Low-fidelity prototyping
- Storyboarding.

[20 minutes]

4 **Prepare key questions** for participants to use throughout the workshop.
E.g. 'What do you currently enjoy/not enjoy about shopping?'
E.g. 'What would an ideal shopping experience look like for you?'
E.g. 'What are some features of this design that you like?'
E.g. 'What would you change?'
[10 minutes]

5 **Run the workshop.** Be sure to communicate the purpose and intended outcomes. Explain the purpose of the design, without too much detail, as this can limit the creativity of the participants. Introduce each activity as it starts. **Allow participants to design concepts and augment existing ideas** with their suggestions. Offer templates and frameworks to assist participants with completing the chosen methods.
[1–4 hours]

6 After the workshop you can **interpret the collected data** using affinity diagramming (p.22) or thematic analysis (p.122). Gather the feedback and concepts from the co-designers. How does this influence the design concept?

Competitor Analysis

Knowing how you compare to those around you

ACADEMIC RESOURCES:

Nusem, E., Wrigley, C., & Matthews, J. (2015). Exploring aged care business models: a typological study. Ageing & Society, 1-24.

O'Shaughnessy, J. (1995). Competitive marketing: a strategic approach. Routledge.

The competitor analysis method is a flexible approach for understanding where a product or service fits compared to offers available in the market. The method can be used to evaluate existing products or services against the market, or to identify opportunities for a new product or service.

The initial step is to identify the market. Depending on the reach of the envisioned design solution, this can involve either local competitors (e.g. in the same city or the same country) or global players. Even if the design solution is only addressing a local market, it can still be valuable to also consider a few international competitors.

The method relies on a 'coding structure' in the form of a set of variables that is either predetermined or created for the design project. By using this coding structure to collect information about each of the competitors, it is possible to easily compare, monitor and understand the market, and how a new product or service would fit in. A competitor analysis can be complemented with perceptual maps (p.96) by choosing two specific variables and asking potential customers or users to rank each competing product or service.

Completing a competitor analysis early on in a design process is critical to avoid being blindsided by the competition. Assessing the strengths and weaknesses of competitors leads to a better understanding of opportunities to improve competitive advantage. However, any gaps identified in the market need to be carefully considered. Sometimes, there is a good reason why an approach hasn't been implemented by any of the competitors.

EXERCISE

YOU WILL NEED
Internet access, a partner

In this exercise, you will perform a competitor analysis by collecting data and analysing your position in the market using the provided template (p.171). Focus on a product or service of your choice, or follow the 'Designing Space Travel' brief (p.141).

1 **Generate a list of potential competitors** that offer a similar product or service. Consider both local and international companies. Try to identify the top four most significant competitors.
E.g. Expedia's online flight booking portal
[5 minutes]

2 **Brainstorm variables and write them down** in a list. These are relevant factors against which each competitor could be evaluated. Variables may be basic, like price or quality, or be more specific to your chosen sector.
E.g. product categories, customer segments, primary revenue stream
[10 minutes]

3 Review and refine your list to **identify the variables most relevant** to your chosen sector. Write these down in the left column of the competitor analysis table (p.171).
[5 minutes]

4 **Plot each of the four companies** into the competitor analysis table (p.171). Now you can begin your analysis. Are all companies targeting the same customer? Offering the same value? Where do they differ most significantly?
[10 minutes]

5 **Discuss with your partner the strengths and weaknesses** of the competitors, highlighting any gaps or opportunities your company could benefit from.
[10 minutes]

Design. Think. Make. Break. Repeat.

Contextual Observation

Observing how people act in the wild

ACADEMIC RESOURCES:

Goodman, E., Kuniavsky, M., & Moed, A. (2012). Chapter 9. Field Visits: Learning from Observation. In Observing the User Experience: A Practitioner's Guide to User Research (pp.211-238). MA, USA: Elsevier.

How to Conduct User Observations. (2017, February 4). The Interaction Design Foundation. Retrieved from https://www.interaction-design.org/literature/article/how-to-conduct-user-observations

Contextual observation can be used to study people's behaviour in different environments, such as workplaces, homes, public spaces, and so on. Experiences in real life don't happen in a vacuum, and contextual observation takes into account the range of external factors that can influence people's behaviour, such as environmental, temporal and social factors.

Data collected through contextual observation includes a user's actions, posture, changes in facial expressions and gaze, and gestures as they perform a specific task related to a product or service. Analysing this data can reveal aspects of behaviour, workflow, and interactions with existing products or services. In contextual observation, data is mainly limited to people's behaviour that is visually accessible, such as their reaction to input from the surrounding environment, social interactions, and so on. Such an environment provides less opportunity to observe genuine interactions between a person and a product, inclusive of how these are influenced by external factors and the context of use.

Contextual observation can be used to develop a better understanding of a design problem or context, as well as to gather feedback about a prototype design. In the latter case, it provides a more natural alternative to usability testing (p.126), which is typically conducted in a structured lab environment.

To ensure that the observation reveals useful data, some preparation is required. Most importantly, the objective of the observation has to be defined before the observation takes place. Other aspects to plan and consider include audience, location, time of the day, and day of the week.

EXERCISE

YOU WILL NEED
A partner, pen, paper, camera

In this exercise, you will perform contextual observation by observing a user's interaction with an existing product. Use the provided template (p.172) to record your observations.

1 **Choose a topic for your study.** If you don't have an existing product or prototype you would like to observe, choose one from the following list:
E.g. a vending machine, an ATM or a parking meter

2 **Prepare for your study** by writing down answers to the following questions:
- What do you expect to learn from your observations? *E.g. the key problems with the current design*
- What kind of data do you need to collect to improve the design?
- *E.g. moments of frustration, signified by user's facial expressions*

[5 minutes]

3 **Choose the setting**.
E.g. a vending machine in a cafeteria or an ATM in a shopping mall
Consider the following factors:
- Is the product designed for indoor or outdoor use?
- Is it used while doing other activities?
- Is it used by one person or multiple people at the same time?

[5 minutes]

4 **Observe people interacting** with the object of evaluation. For each activity a user performs, **note down** what they are doing:
- Facial expressions;
- *E.g. a smile, shaking head*
- Posture of hands, body and head
- Gestures and body language
- Uttered sounds that communicate emotions, such as pleasure, frustration, etc.

Take pictures of the object and its interface. If you are planning to take pictures of users interacting with the object, make sure to seek their consent first. Alternatively, you can get your partner to stage interactions for the purpose of documenting the object.

[10 minutes]

5 **Review your recording and make notes** on the data collection form (p.172).

[10 minutes]

Design. Think. Make. Break. Repeat.

Cultural Probes

Getting to know your users through playful and provocative tasks

ACADEMIC RESOURCES:

Boehner, K., Vertesi, J., Sengers, P., & Dourish, P. (2007). How HCI interprets the probes. In Proceedings of the SIGCHI conference on Human factors in computing systems. ACM.

Gaver, B., Dunne, T., & Pacenti, E. (1999). Cultural probes. Interactions, 6(1), 21-29. ACM.

Gaver, B., Boucher, A., Pennington, S., & Walker, B. (2004). Cultural probes and the value of uncertainty. Interactions, 11(5), 53-56. ACM.

The cultural probes method supports divergent thinking. It relies on playful and provocative tasks mediated by well-crafted items. First introduced by Gaver et al. (1999), cultural probes are physical packets consisting of various items such as maps, disposable cameras, and postcards. The items come with open-ended and provocative tasks for acquiring inspirational responses from a community of participants about their lives, thoughts and values. In its approach, the cultural probes method draws on situationist art practices such as psychogeography and *dérive*. It values uncertainty, play, exploration and subjective interpretation as ways of dealing with the limits of normative scientific approaches to knowing.

The data collected through cultural probes is typically richly textured, highly personal, fragmented, and partial. This makes it better suited to generating an inspirational resource or developing a subjective understanding of an unknown user group rather than being analysed in a structured way. There are many variants of cultural probes to target different domains or desired outcomes (Boehner et al., 2007).

A typical cultural probes process involves creating a kit containing individual probes with engaging tasks, explaining the probe study and the kit contents to the participating users, delivering the kit, getting back the completed probe tasks, and interpreting the returned probes. While carefully crafting the probe items and tasks plays a key role in inspiring the users, making the probes too polished may discourage the users from using them comfortably.

Cultural probes can be used in the early phases of a design process as a self-reporting method, complementing more conventional user research methods such as interviews and questionnaires. They should not be used as a sole lightweight and rapid data collection method.

EXERCISE

YOU WILL NEED
3–4 people, printed maps, small notepads, envelopes or boxes

In this exercise, you will create a basic cultural probe kit, deliver it to a small group of participants and interpret the returned probes. Focus on your own design problem, or follow the 'Supermarket of the Future' design brief (p.143).

1. **Recruit participants**, explaining to them that the study will require a commitment of several hours over a period of one to two weeks.

2. **Design and prepare probes** to get insights about people's experiences and practices around shopping. Prepare three different kinds of tasks:
   - Task one: a supermarket map that they can annotate with their route and activities.
   - Task two: a notepad with provocative questions requiring textual responses.
   - Task three: a photo diary with questions that require photos to answer them.

   [2–3 hours]

3. **Pay attention to the following**, when designing your tasks:
   - Craft each probe item; make the materials accessible and appealing.
   - Use a mix of text, drawings, sketches and photos to explain the tasks.
   - Leave room for users to interpret the tasks in their own way. Use open-ended questions, oblique wording and evocative images.
     E.g. 'Tell us about your favourite time of day to go shopping'
     E.g. 'What would you change in your neighbourhood supermarket?'
   - E.g. 'Take a photo of your favourite aisle, and tell us why it is your favourite.'

4. **Deliver the probes in an envelope or box** to each of your participants. Briefly explain the details of the tasks. Do not give them too detailed instructions, as this can limit the creativity and authenticity of the responses.
   [1 week]

5. **Follow-up with each participant**. Contact them after a few days to check if they have any questions about the tasks. This can improve the response rate and quality.

6. **Interpret the data contained in the probes** to identify interesting points captured and expressed by your participants. Immerse yourself in the data by highlighting or grouping important responses, for example, using the affinity diagramming technique (p.22).
   [2–4 hours]

Design. Think. Make. Break. Repeat.

Decision Matrices

Because making design decisions isn't easy

ACADEMIC RESOURCES:

Pugh, S. (1996). Creating innovative products: Using total design. Addison-Wesley.

Roozenburg, N. F., & Eekels, J. (1995). Chapter 9. In Product design: fundamentals and methods (Vol. 2, pp. 293-316). Chichester: Wiley.

Indecisiveness has always been part of the human condition. Ancient peoples used esoteric methods to try to see into the future, but we now know that there are other ways to make informed choices – by creating a range of options and systematically evaluating them. This is what decision matrices help us to achieve in a design process.

There are many different forms of decision matrices, but all are based on having an initial set of criteria against which to evaluate our choices. Should your design be heavy or light? Is sustainability an important factor? Must it be easy to use? The set of criteria is generated to suit the ambitions of a specific project.

Regarding its physical form, a decision matrix is a table. In the row headings, we have our criteria. In the column headings, we have the concepts we are comparing, which sometimes includes a 'datum' if there is already an existing design or process. Once these are filled, each concept can be ranked based on how well it performs against the specified criteria. This can be done visually by colouring squares such as in a 'Harris Profile' (Roozenburg & Eekels, 1995) or numerically by scoring each concept such as in a 'Pugh Matrix' (Pugh, 1996). A weighting can also be included, so that particularly vital criteria count for more. Importantly the highest score does not necessarily win – the decision matrix is a tool, not a magic wand. The final decision may be to choose one concept but improve its weakest aspects or combine it with another concept that is stronger on other criteria.

EXERCISE

YOU WILL NEED
Pen, paper

In this exercise, you will create a decision matrix to help choose between a range of design concepts. Rate your concepts using the provided template (p.173). Focus on your own design problem, or follow the 'Museum Visitor Experience' brief (p.142) and use the resources on the companion website.

1 **Generate a list of criteria** to rate your concepts against, based on your design requirements and user needs. Criteria will normally be quite specific to your design task and could include:
- Design requirements: fit with the design brief, match with specific user needs
- Physical aspects: size, weight, height, speed, robustness
- Usability: easy to understand, efficient at performing tasks, effective
- Feasibility: cost, time investment, technological constraints

Consider how you might measure these criteria – are they objective or subjective?
[10–20 minutes]

2 **List all of your criteria in the first column** of the decision matrix, in order of priority. You should have at least eight to perform a successful evaluation.
[5 minutes]

3 **Write your design concepts across the top row** of the decision matrix, one in each column. If it is a redesign, the existing situation should be written in the first column, this acts as a 'datum' and allows for direct comparison to new ideas.
[10 minutes]

4 **Add weightings.** Are certain criteria more important to the success of the design than others? These could be weighted more heavily. Each score is multiplied by the weighting factor in these cells of the matrix.
[10 minutes]

5 **Rate each idea according to each criterion.** Give it a number between -3 and +3, according to how successful it is in fulfilling that criterion. Criteria that are weighted more heavily are multiplied.
[10–20 minutes]

6 **Add up the totals**, to see which design concept scores the most highly. You could choose the most high-scoring one. Alternatively, you might use the decision matrix to determine which aspects are weak and need more work, or whether you can combine the best aspects from several concepts to form a new concept.
[10 minutes]

Design. Think. Make. Break. Repeat.

Design by Metaphor

The power of seeing something as something else

ACADEMIC RESOURCES:

Hey, J., Linsey, J., Agogino, A. M., & Wood, K. L. (2008). Analogies and metaphors in creative design. International Journal of Engineering Education, 24(2), 283-294.

Madsen, K. H. (1994). A guide to metaphorical design. Communications of the ACM, 37(12), 57-62.

Schön, D. (1979). Generative Metaphor: A Perspective on Problem-setting in Social Policy. In A. Ortony (Ed.) Metaphor and Thought (pp. 254-283). Cambridge: Cambridge University Press.

In linguistics, metaphors are 'A figure of speech in which a word or phrase is applied to an object or action to which it is not literally applicable' (Oxford English Dictionary). In design, metaphors are used to refer to familiar precedents from the world around us. The metaphor assists the transfer of what we know in one domain (the source) into another different domain (the target).

The desktop metaphor used in the graphical user interface of personal computers is a classic example of using a metaphor to aid conceptual understanding of interactions. Applying metaphors from an office environment allows people to easily learn how to perform interactions within a graphical user interface. Like on a physical desk, files are located inside folders. Unwanted files are dragged into a rubbish bin.

Applying different metaphors during the conceptual design phase can reveal a variety of design solutions. For example, the metaphor of 'computers as humans' leads to dialogue-based forms of interaction, whereas the metaphor of 'computers as tools' leads to direct manipulation-based forms of interaction. Approaches to design that are modelled on nature are referred to as biomimicry. Far-fetched analogies can be used to spark ideas.

Metaphors can also be used to generate new perspectives on a problem by seeing something as something else (Schön, 1979). Metaphors can be applied in design to explore the problem domain from unusual and competing perspectives, and to reveal the hidden dimensions of a problem to be solved (Madsen, 1994). By applying a series of strategic questions, it is possible to unpack the metaphor and discover a new understanding of and potential solutions to the problem at hand.

EXERCISE

YOU WILL NEED
A partner, pen, paper

In this exercise, you will explore a problem domain through the application of several metaphors. Use the provided template (p.174) to get you started. Focus on your own design problem, or choose a design brief (p.138).

1 **Choose two very different metaphors** for exploring a problem domain. Each metaphor should be familiar to you and your partner but complex enough to generate interesting perspectives.
E.g. a relay race
E.g. a swarm of bees
[5 minutes]

2 **Explore the problem space** by using the following questions to provoke thinking. Focus on the first metaphor:
- **Tell the metaphor's story.** Talk about the problem domain as if it was the chosen metaphor.
- **Elaborate the triggering concept.** The triggering concept may either be a key concept in the source or the target domain. The concept is elaborated upon to cover activities that are not normally understood in those terms.
- **Look for new meanings for the concept.** Metaphors suggest new meanings for existing concepts.
- **Elaborate assumptions.** Make explicit what the metaphor hides and what it highlights.
- **Identify the unused part of the metaphor.** Look for unused aspects, features and properties in the source domain and consider how they may play a role in the target domain.

[20 minutes]

3 **Switch to the second metaphor** and repeat the above process.
[20 minutes]

4 **Compare and discuss** with your partner what you learnt from exploring the problem domain using the two metaphors.
[5 minutes]

5 **Select one of the metaphors as a potential model** for a design solution. What properties, features and relationships of the metaphor will you transfer to your design solution? Sketch and annotate the first draft of your design concept.
[10 minutes]

6 Can you **extend the metaphor** beyond a literal application? Consider the strategic questions you explored above to help generate new, unexpected or interesting interpretations of the metaphor. Sketch and annotate the second draft of your design concept.
[10 minutes]

Design. Think. Make. Break. Repeat.

Design Critique

Valuing the perspective of others

ACADEMIC RESOURCES:

Kolko, J. (2011). Endless nights-learning from design studio critique. Interactions, 18(2), 80-81. ACM.

The value of the design critique method is three-fold. First, it yields valuable feedback quickly, regularly and at low cost without having to involve participants from the target audience, as is the case, for example, in usability testing (p.126). Second, it allows people to develop their skills in providing constructive feedback and in doing so to improve their own design knowledge. Third, it builds resilience in the designer to be open to criticism of their work, and thus improve its quality.

Design critiques (also called 'design crits') focus on evaluating existing ideas rather than coming up with new ones. Design critiques ideally involve three to seven people but can also be done in groups of two. Participants are typically other design team members, although it can also be of value to include other project stakeholders.

Typically, design critiques take place in a studio environment, where people can present their design on a wall or using a digital screen or projector. While one person presents their design, the other participants take notes. Learning how to give and take criticism in a respectful and generous way will also help to build trust within the design team.

Depending on the project, design critiques may be conducted on a daily or weekly basis. It is important to keep the presentation short, and to include a brief summary of the project and user requirements progress since the last critique session as well as any points that require feedback or clarification. The material used in the presentation should include sufficient detail to allow participants to criticise overall themes as well as specific details.

EXERCISE

YOU WILL NEED
2–10 people, Post-it notes (3 colours), marker

In this exercise, you will critique design work using three simple lead-in statements. You will practice communicating critique in a respectful and constructive way. Focus on your own design problem, or use the resources on the companion website.

1. **Present your project to your peers** using design artefacts such as affinity diagrams, storyboards, personas, wireframes, concept diagrams, etc.
[5-10 minutes]

2. While you present, your peers **respond to the statement 'I didn't know that …'** and write something they learnt from the presentation on a **blue** Post-it note.
E.g. I didn't know that most people who walk through this place listen to music on headphones.

3. While you present, your peers **respond to the statement 'Tell me more about …'** and write something related to the work that they want to know more about on a **yellow** Post-it note.
E.g. tell me more about what happens when it gets dark.

4. While you present, your peers **respond to the statement 'Have you thought about …'** and write something they think is missing from the work (but might be useful) on a **pink** Post-it note.
E.g. have you thought about asking similar questions to teenagers?

5. When recording feedback, keep the **following guidelines**:
   - Feedback should focus on how/why a design does or does not satisfy a user need.
   - Reviewers should ask clarifying questions when necessary.
   - Presenters should answer questions without getting defensive.
   - You can use the words 'I like…' and 'I don't like…' but try to refer back to user needs.
   - Avoid problem-solving. The focus is on analysing the proposed solution, not suggesting other approaches.

   Give your final set of Post-it notes to your peer when the critique is finished.

6. **Review the Post-it notes you received and discuss any questions** you might have. Reflect on the value of the feedback received.
[5-10 minutes]

7. **Swap roles** until everyone in the peer critique session has had a turn.
[5-10 minutes per team member]

Design. Think. Make. Break. Repeat.

Direct Experience Storyboards

Acting out user experiences when involving users is not possible

ACADEMIC RESOURCES:

McQuaid, H. L. Goel, A. & McManus, M. (2003, June). When you can't talk to customers: using storyboards and narratives to elicit empathy for users. In Proceedings of the 2003 international conference on Designing pleasurable products and interfaces (pp. 120-125). ACM.

To understand user needs, it is important to examine people's practices in detail, including the environment in which they use services. Direct experience storyboards is a method that combines the advantages of systematic observation, direct experience, documentation and storytelling. The method is useful for contextually sensitive situations where it may not be possible or convenient to conduct direct-user research, such as hospitals, libraries, museums and other public and private institutions with similar characteristics. By acting out a given scenario in an authentic setting, designers can capture both positive and negative aspects of a user's experience, and collect impressions of what it is like to interact with a product or service.

The direct experience storyboard method involves the following activities: observing people and their interactions with products or services, creating a list of typical tasks they perform, assigning each task to a design team member, getting each team member to perform their tasks while taking photos or screenshots, annotating the printed photos with specific details of the associated experience, and creating a photo-narrative storyboard by hanging the annotated photos in chronological order. A photo-narrative storyboard is a kind of storyboard with some photographs of the same scene replacing the hand-drawn sketches of a conventional storyboard (p.120).

After creating the storyboards, team members come together and discuss the experiences depicted in the storyboard. This method is particularly useful for understanding user experience from a direct first person experiential viewpoint, in an authentic setting. The storyboard and the discussion around the captured experiences can inform the construction of a concept map or information architecture involving various components of a future system and interactions around them.

EXERCISE

YOU WILL NEED
3–4 people, smartphone, pen, paper, sticky tape, acetate paper (optional)

In this exercise, you will visit a location and re-enact a typical user activity. You will generate a photo-narrative storyboard that captures your direct experience of this activity. Focus on your own design problem, or follow the 'Museum Visitor Experience' brief (p.142).

1. **Visit a location** that is relevant to your design brief, and observe some of the typical activities or tasks that users perform. Take notes and photographs to record your observations of how they perform these activities.
E.g. observe other visitors and museum staff.
[1 hour]

2. **Create a list of these typical activities.** They may include things like:
E.g. orienting and navigating around the building
E.g. searching for a particular object in the collection
E.g. finding out more information about an interesting topic
Include brief descriptions of each activity that you have observed.
[10 minutes].

3. **Assign one or more activities to each team member.** Ask each team member to perform or role-play their assigned activities from start to finish. While carrying out these activities, take photos or screenshots to record all the key steps.
[10–20 minutes]

4. **Print out your photos and annotate them.** Sort them into chronological order and stick them to a piece of paper to create the storyboard. You can annotate your photos directly or on an acetate overlay.
[10 minutes]

5. **Discuss your direct experience storyboard with your team.** Reflect on the positive and negative aspects of your experience. Where is there room for improvement or optimisation? What aspects should stay the same?
[10 minutes]

Design. Think. Make. Break. Repeat.

Empathic Modelling

Putting yourself in someone else's shoes

ACADEMIC RESOURCES:

Fulton Suri, J., Battarbee, K., & Koskinen, I. (2005, April). Designing in the dark: Empathic exercises to inspire design for our non-visual senses. In Proceedings of International conference on inclusive design (pp. 5-8).

Kullman, K. (2016). Prototyping bodies: a post-phenomenology of wearable simulations. Design Studies, 47, 73-90.

McDonagh, D., & Thomas, J. (2010). Rethinking design thinking: Empathy supporting innovation. Australasian Medical Journal, 3(8), 458-464.

Nicolle, C. A., & Maguire, M. (2003). Empathic modelling in teaching design for all.

When designing products or services, it is critical to ensure that they are accessible to all users. Empathic modelling is a method that prompts us to think beyond designing for an ideal user of an ideal height with ideal eyesight and ideal motor skills. It is used to simulate, from a first person perspective, some of the everyday challenges people with reduced physical or perceptual abilities experience. This allows designers to develop an empathic connection and use other people's experiences to inform the design of products or services.

As with other techniques based on role-playing, empathic modelling requires active, physical engagement with the task. It is generally assisted by external elements such as suits, props or virtual reality headsets that simulate someone else's perspective. These tools are intended to help designers replicate what it is like to experience a world perceived through someone else's physical capabilities. Common tools used in this method include blurry glasses, blindfolds and wheelchairs. In some cases, bodysuits are used to simulate specific health conditions, such as Ford's 'Third Age Suit', which restricts the movement of arm, torso and leg joints, and includes gloves and tinted glasses. However, empathic modelling can also be achieved by using simple, readily available materials and props.

Ultimately, representatives of user groups reduced physical or perceptual abilities should be consulted as part of the design process, but the strengths of this method are its applicability at an early stage without requiring access to participants and the first person experience that it delivers to members of the design team.

EXERCISE

YOU WILL NEED
A partner, camera, cling wrap, belt

In this exercise, you will put yourself in the shoes of someone with reduced sensory or physical capabilities. In order to do so, you will use some simple materials to alter your senses. Use the resources provided on the companion website to create your own glasses prop, or use readily available materials.

1. Preparation: Cover your mobile phone camera lens with a crumpled piece of cling wrap. Take a test photo to make sure you get a blurry shot. Cover the glasses with crumpled cling wrap – you should still be able to see a little bit.
[5 minutes]

2. Visit a place where you have been before, together with your partner, for example, the building where you study or work.

3. With the help of your partner, **explore this familiar place from an unfamiliar perspective**. Move slowly, recognise, stop, take photos of things that catch your attention. Focus on properties that you may not have paid attention to in the past, such as colours, ambiguous shapes, and sounds.
[15 minutes]

4. Switch roles with your partner and help them carry out the same exercise, assisting them to walk safely.
[15 minutes]

5. Discuss your experience with your partner, using the following questions:
- Which kind of environmental information was revealed after experiencing a familiar place from an unfamiliar perspective? Your photos might help you to reflect on this.
- How much can we infer about our location when we rely on sound or colour?
- How difficult can it be to rely on our sense of orientation? Imagine yourself having reduced physical capabilities in another situation such as taking public transport. Does public transport rely on perceptual cues beyond sight?
- How could such a scenario be improved with more inclusive design?

[10 minutes]

6. Try the same exercise to simulate the experience of someone with reduced mobility. Restrict the movement of your dominant arm using a belt, scarf or bandage. Perform everyday activities such as eating, typing or pouring water, and reflect on the challenges and insights that emerge.
[15 minutes]

Experience Prototyping

Turning ideas into something that can be experienced

ACADEMIC RESOURCES:

Buchenau, M., & Suri, J. F. (2000, August). Experience prototyping. In Proceedings of the 3rd conference on Designing interactive systems: processes, practices, methods, and techniques (pp. 424-433). ACM.

Prototypes typically explore the tangible qualities of a design solution. However, they can also be used to test and explore the intangible qualities of a design. The experience prototyping method enables a focus on the user experience and context of use, rather than fixating on the product form and features. It highlights the experiential dimensions – how people think, feel and act in situations of use with future products and services. Experience prototyping is commonly used when the physical, social, spatial and temporal dimensions of a situation are important to explore.

Experience prototyping is closely related to low-fidelity prototyping (p.84), as it involves the use of low-fidelity representations of products and environments. The focus is less on the fidelity of the physical prototype and more on the simulation of the experience made available through the form of the prototype. Typically, life-size mock-ups of the system and space are created using the materials at hand. The experience is brought to life by bodystorming (p.26) and role-playing (p.108) scenarios of interest in the setting of the physical prototype. Once an experience prototype has been created, it can be used to spot flaws in a concept and discover opportunities.

Typically experience prototyping is performed by and with members of the design team, but it can also be used to test out the experience with users or other stakeholders by letting them try it out. By testing designs in a physically active way before thinking about the details, it is possible to gain a thorough appreciation of the future experiences that are being created.

EXERCISE

YOU WILL NEED
3+ people, sticky tape, paper, cardboard, pins, markers, furniture, Post-its

In this exercise, you will construct a life-size physical prototype that highlights key aspects of the experience you are designing for a future product or service. Use your finished prototype to try out and evaluate your design.

1 **Select a problem or situation** to redesign.
E.g. how can you improve the experience of buying a cup of coffee?
Start by **thinking about the process of using** your product or service. Write down the five most crucial steps in this process.
E.g. enter store and join queue --> place order with cashier --> pay --> barista makes coffee --> collect coffee
[10 minutes]

2 **List the qualities of experience** that you think are important for each step from the perspective of the user.
E.g. a regular customer may like to be recognised and greeted by name, bringing qualities of friendliness to the interaction.
Think about how you can **translate** these qualities into the experience prototype.
[10 minutes]

3 **Think about which aspects** of this experience may be problematic, innovative or important to test with users. **Make these the focus** of your prototype. Take notes in your sketchbook to plan the creation of your experience prototype.
[10 minutes]

4 **Build your experience prototype** using the available materials, space and your group members. Remember, the focus is on communicating key aspects of the envisioned experience of your design to potential users, stakeholders or other designers.
E.g. for the coffee example, the experience prototype could be built as follows:
- *Indicate which part of the room is the store using signage and tape.*
- *Rearrange tables to represent the counter; use paper signs to indicate zones.*
- *Use team members to play the cashier and barista; use Post-it notes to identify roles.*
- *Make paper drawings of screens to represent a digital payment system.*
[60 minutes]

5 **Walk through** the finished experience prototype yourself in the 'shoes' of a customer. Note down any parts of the system that cause frustrations or present opportunities.
[10–20 minutes]

Design. Think. Make. Break. Repeat.

Experience Sampling

Sampling people's states, emotions and thoughts in real-time

ACADEMIC RESOURCES:

Larson, R., & Csikszentmihalyi, M. (1983). The experience sampling method. New Directions for Methodology of Social and Behavioral Science, 15, 41-56.

The experience sampling method (ESM) allows collecting real-time self-reporting data about subjective experiences such as mental state, emotions or thoughts. The method captures such experiences by asking respondents to stop at predetermined times to report on their state or how they are feeling. The time of reporting is either predetermined based on events or signalled, for example, using mobile phone text messages, at which point the respondent will stop what they are doing and answer a questionnaire or report their experience in a diary.

Asking respondents to report in the moment has the advantage of being less reliant on their memory when reporting experiences. The method can, therefore, eliminate bias that often occurs in retrospective self-reporting of events and increase the validity of the results collected from participants. By repeatedly recording a participant's experiences over a specific period, the dynamics of changes in their experience (e.g. changes in emotion) are uncovered. The depth of data generated can bring to light opportunities for new design solutions or improving the existing solutions. Data collected through ESM can be analysed using either quantitative or qualitative methods. It is also possible to use the recorded data during an 'exit' interview, i.e. an interview conducted at the end of the study. By showing the data to the participant, they are able to refer back to the moment when they recorded that data, rather than relying on memory alone. We can then record additional insights from a participant's recall of that moment.

One challenge in administering ESM it requires repeated self-reports, making it a time-consuming data collection technique. Additionally, it is important to create tasks that are feasible for participants, so they stay motivated and provide complete reports.

EXERCISE

YOU WILL NEED
3–4 people, 2–5 participants, pen, paper

In this exercise, you will collect data in situations where changes in momentary experience are important. This offers an understanding of contextual factors that influence your users' needs, expectations or experiences with a product or service. Focus on your own design problem, or choose a design brief (p.138).

1 **Invite participants** to engage in an experience sampling study for three days.
E.g. their daily commute to work by public transportation

2 **Prepare a questionnaire** (p.102) that incorporates questions about your participants' emotional responses to the environment and people around them. Relate each question to a specific moment:
E.g. reporting how they feel immediately before they start their commute
E.g. reporting how they feel right before leaving the public transportation
Alternatively, you can use the ESM questionnaire provided in the resources on the companion website.
[60 minutes]

3 **Create a hard copy booklet** of all the questionnaires, with sufficient space for the respondent to specify the time and date of each report.
[30–60 minutes]

4 Distribute the booklet to your participants and ask them to **record their momentary experiences**. They should be instructed to:
- Report honestly, describe momentary experiences as they happen
- Describe each experience individually, try not to compare new experiences to previous ones
- Try not to actively look for things to report
- Note down the reasons for any ratings, frustrations and good experiences.
E.g. I am feeling amused because a child is playing hide-and-seek with me.

5 Collect the booklet after three days and **analyse the comments**, e.g. using thematic analysis (p.122). Determine the average of rating scales. What is this data telling you? Are there any opportunities for improving the experience?
[30–60 minutes]

Extreme Characters

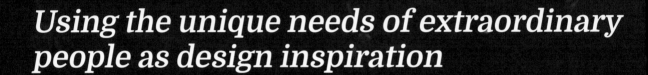

Using the unique needs of extraordinary people as design inspiration

ACADEMIC RESOURCES:

Bell, G., Blythe, M., & Sengers, P. (2005). Making by making strange: Defamiliarization and the design of domestic technologies. ACM Transactions on Computer-Human Interaction (TOCHI), 12(2), 149-173.

Djajadiningrat, J. P., Gaver, W. W., & Fres, J. W. (2000). Interaction relabelling and extreme characters: methods for exploring aesthetic interactions. In Proceedings of the Designing interactive systems: processes, practices, methods, and techniques (pp. 66-71). ACM.

In a typical design process, it is common to focus on a particular target group of users and spend time understanding their problems, needs and motivations. Although this is an important step in a design process, the insights from this well-defined group may be limited to a small set of emotions and practices. The method of extreme characters allows going beyond considering the needs of typical and conventional users, by prompting us to think about design solutions for extreme characters, such as a drug dealer or a secret service agent. It encourages divergent thinking through 'defamiliarisation' and by moving outside of well-defined problem spaces to access a larger spectrum of human emotions and practices.

Extreme characters, with their unusual and unique emotions, habits, and needs, allow us to expand our set of concerns and to discover new possibilities for design solutions. The method involves imagining extreme characters and their character traits, values, and typical daily practices. Once an extreme character has been formed, it can then be used to come up with design ideas that satisfy their unique needs. The rationale behind this method is that some of these design ideas can then be fed back into the design process, for example, by proposing a new feature that could also be useful to typical users.

The extreme characters method can be used in the early phases of a design project to identify new aspects of a problem and to generate new design ideas. Using supplementary visuals, such as photographic representations of the extreme character and their surroundings, and bodystorming (p.26) or role-playing (p.108) with other group members can help to create a more complete character depiction.

EXERCISE

YOU WILL NEED
3–4 people, pen, paper, Post-its

In this exercise, you will create an extreme character, record their details using the provided template (p.175), and use the character to generate design ideas. Focus on your own design problem, or follow the 'Autonomous Vehicles' brief (p.140).

1 In your group, **choose an extreme user type** you would like to pursue. Try to find characters that fit the context of the chosen design brief, such as:
- The good Samaritan (focused on community and culture)
- The romantic (focused on dreams and pleasure)
- The Trekkie (focused on sci-fi and high-tech)
- The money-spinner (focused on wheeling and dealing).

Have a quick discussion about this extreme user type with your group.
[5 minutes]

2 **Compose a character** for the extreme user type that your group chose. Each individual writes down their character's distinct features using the template included in the resources:
- Who they are?
- What do they do?
- What do they value?
- What do they need?
- What motivates them?

Add sketches or visuals to enhance your description.
[5 minutes]

3 **Create an attitude for each character.** Write down each character's typical practices, choices and values in relation to the chosen design problem. Focus on the aspects of the characters that are potentially relevant.
[5 minutes]

4 **Act from your character's point of view** during an idea generation session. Generate design ideas for your extreme character in relation to the design problem and record each idea on a Post-it note.
[15 minutes]

5 **To explore your character's behaviour** and interactions with the proposed design solution further, consider writing a scenario (p.110) and creating a storyboard (p.120).

Focus Groups

Gaining insights by observing and listening to group discussions

ACADEMIC RESOURCES:

Courage, C., & Baxter, K. (2006). Understanding your users: A practical guide to user requirements methods, tools, and techniques. San Francisco: Morgan Kaufmann.

Tremblay, M. C., Hevner, A. R., & Berndt, D. J. (2010). The use of focus groups in design science research. In Design Research in Information Systems (pp. 121-143). Springer US.

As its name suggests, a focus group involves a group of people having a focused discussion on a specific topic. Like interviews (p.78), focus groups elicit understandings through asking questions and prompting conversations. While interviews typically focus on just the interviewer and the interviewee, a focus group allows for findings to emerge from the social dynamic of a group discussion. Focus groups allow designers to gather information across many phases of the design process. They are most useful for gathering concrete information about existing experiences, building empathy or understanding with users, defining the design problem, and providing feedback on design concepts and prototypes.

Because focus groups can include a large number of people (e.g. 6–15 participants), they can be a quick and easy way of gathering a lot of data. Focus groups allow you to gauge the qualitative opinions, attitudes, practices, needs and priorities of users, experts or stakeholders that you may not know very much about. Their group nature makes them especially relevant to designers working with different target groups. For example, in evaluating one system, a focus group with novice users may reveal vastly different things to one conducted with expert users. Depending on the social dynamic of the specific focus group, participants may vocalise or withhold honest information.

Focus groups typically adopt an open and conversational structure that allows designers to investigate topic areas in depth. A moderator helps guide the group through setting the agenda, asking questions, and prompting participants for further detail. An observer or scribe may also be part of the facilitation. While a focus group must always be focused on a specific topic, it is still important to let findings emerge naturally from the discussion, and for participants to feel free to explore interesting tangents.

EXERCISE

YOU WILL NEED
A partner, 4–6 participants, pen, paper, audio/video recorder (optional)

In this exercise, you will collect real data by conducting a short focus group. Follow the basic script, but pursue interesting leads by keeping the conversation flowing. Use provided template (p.176) to take notes. Focus on your own design problem, or follow the 'Supermarket of the Future' brief (p.143).

1 **Draft the structure of your focus group.** Delegate the following roles and responsibilities with your team:
- Moderator: responsible for running the focus group and keeping people talking.
- Scribe/observer: responsible for taking notes on what participants say and do.

[10 minutes]

2 **Prepare a space for 4–6 participants.** Create a quiet and private setting to limit distractions for your participants. The seating is often a circular arrangement with the moderator positioned in the middle and the scribe on the side.
You may wish to set up audio or video recording equipment to record your session.

[20 minutes]

3 **Introduce your study and warm-up your participants.** The moderator should ask everyone to introduce themselves and include some relevant background information.
E.g. your name, what you do, where you shop?
The scribe should record this information. As this is a group environment, be careful about asking for sensitive information such as age or weight.

[5 minutes]

4 **Use the following open-ended questions** to get the conversation started. These can be tailored to your topic:
- Tell us about the last time you went to …
- What were the best parts of the experience?
- What caused you frustration or prevented you from doing something?
- What would you like to change about the experience?

Allow the participants to follow relevant tangents, but try to make sure everyone gets the chance to speak at some point.

[15 minutes]

5 **Conclude your focus group** by thanking your participants for their time.

6 With your partner **discuss your findings** – what have you learnt?
- What key findings answer your original questions?
- What understanding of the design problem do you have now that you didn't have before?
- How can this information be translated into new design proposals?

Design. Think. Make. Break. Repeat.

Forced Association

Unlikely matches lead to new ideas

ACADEMIC RESOURCES:

Kokotovich, V. (2004). Non-Hierarchical mind mapping, intuitive leapfrogging, and the matrix: tools for a three phase process of problem solving in industrial design. In DS 33: Proceedings of E&PDE 2004, the 7th International Conference on Engineering and Product Design Education, Delft, the Netherlands, 02.-03.09. 2004.

McFadzean, E. (2000). Techniques to enhance creative thinking. Team performance management: an international journal, 6(3/4), 62-72.

Often in idea generation there comes a point where all the obvious ideas have been explored and the designer gets stuck; coming up with additional ideas becomes an uphill slog. At this point, we can turn to lateral thinking techniques to help shake us loose from our patterns of thinking. Forced association is one of these lateral thinking techniques.

When using forced association for idea generation, we combine disparate concepts and try to come up with ideas based around them. The forced association exercise starts with a list of objects or themes that are related to the problem or brief that you are working on. This list could be a set of relevant keywords such as user goals, possible activities, types of target users or potential technologies that could help solve the problem or brief.

The method can also be carried out by using an existing mind map about the topic and randomly selecting keywords from it (Kotokovich, 2004).

After selecting two or more keywords at random, the next step is to generate an idea for a solution that includes all of the things described by the keywords. Sometimes this might feel difficult or illogical, which is why we call these associations 'forced'. By forcing difficult or illogical associations, it is possible to generate ideas that are different from those you would produce using brainstorming or other idea generation techniques.

The process can be repeated as often as you want; once you have exhausted one combination a new set of mismatched words is drawn – and a new idea generated.

EXERCISE

YOU WILL NEED
Pen, paper

In this exercise, you will generate out-of-the-box ideas using a set of keywords that are 'forced' together in unlikely combinations. Focus on your own design problem, or follow the 'Museum Visitor Experience' brief (p.142), and the resources on the companion website.

1. **Start by making a mind map** about a topic closely related to your design brief. You can use the mindmapping technique (p.88). If following the 'Museum visitor experience' brief, you can skip this step.
   [10–15 minutes]

2. **Choose two keywords at random.** If using the cards, select two different coloured cards. If using a mind map, point two fingers at the page with your eyes shut (no cheating!). These two words will become the stimulus for generating ideas.

3. **Write or sketch down one to three ideas** for how to tackle your design brief in a way that incorporates both of these keywords, or is inspired by them. Even if the task seems difficult try to come up with at least one idea – these are after all 'forced' associations.
   [10–15 minutes]

4. **Repeat the exercise** with a different selection of keywords from your mind map or list, as many times as you like. The ideas you produce can be used directly, or as inspiration for more feasible concepts.

Design. Think. Make. Break. Repeat.

Future Workshops

Collectively envisioning future solutions

ACADEMIC RESOURCES:

Kensing, F. and Madsen, K. H. (1992). Generating visions: future workshops and metaphorical design. In Greenbaum, J. and Kyng, M.,editors, Design at Work: Cooperative Design of Computer Systems, pp. 155–168. L. Erlbaum, Hillsdale, NJ.

Lauttamäki, V. (2014). Practical guide for facilitating a futures workshop. Finland Futures Research Centre. Turku School of Economics, University of Turku.

Workshops, one of the fundamental methods of participatory design, bring together a group of people consisting of designers, researchers, users, client, and other stakeholders sharing a common interest to solve a problem. Workshops foster empowerment, democracy, teamwork, and collective thinking and making. Designers usually work as a facilitator in workshops. The future workshops method allows us to identify the problems related to a current situation, imagine alternative solutions, and outline an action plan for implementing these solutions. It is a structured workshop with three phases: critique, fantasy and implementation.

In the critique phase, a critical understanding of the current situation is developed. Participants brainstorm and identify problems and write down brief statements, observations and critiques about them. The written problem areas are then clustered in the form of a mind map.

In the fantasy phase, participants get creative and imagine various possible solutions for the problems without considering any limitations or restrictions. The focus is upon 'what-if' scenarios, employing other techniques, such as metaphors (p.50). For example, in a library project, metaphors such as the 'library as warehouse' or the 'library as a store' can be introduced to the participants. These metaphors serve as conceptual and inspirational devices designed to help the participants to come up with out of the box solutions (Kensing and Madsen, 1992).

In the implementation phase, the proposed solutions are evaluated from a more realistic perspective. Ultimately, this phase aims to produce a draft plan identifying the resources required for the desired changes: what needs to be done, by whom, with what resources, when, and so on?

EXERCISE

YOU WILL NEED

3–6 participants, A3 or A2 paper, pens, coloured markers, Post-its

In this exercise, you will facilitate a future workshop involving a group of people in order to understand the problems related to a specific problem area. Focus on your own design problem, or follow the 'Autonomous Vehicles' brief (p.140) and envision future solutions. Use the templates the companion website to guide you.

1 **Plan your workshop** by filling in the template. Decide on which other methods you plan to use with your participants during the workshop.
[15–30 minutes]

2 **Gather your participants** around a table and **explain the workshop process.**
[5 minutes]

3 **Conduct a critique session** where the participants brainstorm the problems they observed or experienced in relation to the design brief. Methods such as brainwriting 6-3-5 (p.28) usually lead to better results than conventional brainstorming. Use Post-it notes to write down problems and comments. Create a mind map (p.88) of the identified problems to reveal problem areas.
[20 minutes]

4 **Conduct a fantasy session** where the participants envision as many solutions as possible for the problem areas identified in the previous session. Do not consider any constraints at this stage. Encourage participants to build on each other's ideas. Employ what-if scenarios to help come up with ideas.
E.g. what if there were no roads in your neighbourhood?
E.g. what if car parks in city centres were no longer needed?
E.g. what if your car were like an office?
Use Post-it notes to record the ideas.
[30 minutes]

5 **Conduct an implementation session** where the participants discuss and evaluate the ideas realistically. Produce a draft plan with actors, resources, and a timeline for the most promising ideas.
[30 minutes]

Design. Think. Make. Break. Repeat.

Group Passing

Collaboratively coming up with new ideas

ACADEMIC RESOURCES:

Van Der Lugt, R. (2002). Brainsketching and How it Differs from Brainstorming. Creativity and Innovation Management 11(1), 43-54.

The group passing technique – also referred to as 'brainsketching' technique – is a form of brainstorming that engages a team of people in collaboratively producing new ideas, solutions or design concepts. By working together, the team can draw on each other's creative inputs. The technique seeks new perspectives on each idea from all members of the team, encouraging them to contribute to an idea while avoiding criticism or negative feedback.

In the first step, each team member records an idea on a piece of paper as either sketches, written text or a combination of both. In the second step, each member passes the paper to the person on their left and develops the original idea further by adding more details to it. Papers are circulated around the group until everyone has contributed to everyone else's idea. Once the passing of the papers is complete, and each member receives their original paper back, the ideas are read out aloud followed by a discussion of features, details or solutions contributed by other members.

Like other brainstorming techniques, group passing can be used in the ideation stage of a design project. Beyond being a method for ideation, it is also a powerful technique for fostering teamwork and collaboration. It facilitates constructive collaboration and helps to surpass potential anxiety about sharing incomplete ideas within a group.

While group passing results in a smaller number of outputs compared to other brainstorming techniques, the outcomes are more elaborate in content. In addition, the freestyle writing and drawing of ideas eliminate constraints for expressing new and out of the box ideas.

EXERCISE

YOU WILL NEED
3+ people, pen, A4 paper

In this exercise, you will use the group passing technique to collaboratively generate a high number of design solutions. In this technique you build upon each other's ideas, using annotated sketches. Focus on your own design problem, or choose a design brief (p.138).

1 **Write the topic** for your brainstorming session at the top of your sheet of paper. Discuss the topic in your group, sharing any pre-existing knowledge and any research that you may already have collected.
[5 minutes]

2 **Sketch an idea** that responds to the design problem. Add details about the context, environment or user if this helps to communicate your idea.
[5 minutes]

3 Each member of the group takes turns to briefly **explain their idea**. Avoid asking questions or giving feedback to others at this stage.
[5–10 minutes]

4 **Pass your paper** to the person sitting on your left, then begin the next round.

5 **Examine the idea that you have been given**. Think about how you can improve it, develop it or add more details to it. These could be new features, new forms, details about the interaction with user(s), aesthetics, etc. If a new idea is triggered by the initial idea, try sketching a combination or noting down their differences.
[5 minutes]

6 **Each member explains their additions to the idea** that they were given. Take turns until everyone has presented their ideas.
[5–10 minutes]

7 **Repeat the process** until everyone has contributed to each of the ideas, and each piece of paper has been returned to its original owner.
[5 minutes per cycle]

8 **Take note of the new additions** to your original idea. How has it evolved? Discuss the merits of the final ideas with the group.
[10 minutes]

Design. Think. Make. Break. Repeat.

Hero Stories

Envisioning ideas through speculative storytelling

ACADEMIC RESOURCES:

Dickey, M. D. (2006). Game design narrative for learning: Appropriating adventure game design narrative devices and techniques for the design of interactive learning environments. Educational Technology Research and Development, 54(3), 245-263.

Hinyard, L. J. & Kreuter, M. W. (2007). Using narrative communication as a tool for health behavior change: a conceptual, theoretical, and empirical overview. Health Education & Behavior, 34(5), 777-792.

The hero stories method involves the creation and evaluation of speculative stories around key experiences that people might have with an envisioned product or service. Typically, a hero story focuses on a single user's experience, for example, based on a previously developed persona (p.100). Different to persona-based walkthroughs (p.98), which focus on common tasks or scenarios, hero stories explore extreme scenarios.

Using storytelling techniques, a hero story is developed starting with an ordinary person facing a significant challenge. As the story unfolds, the person finds ways to overcome this challenge. Hero stories are developed using a structured framework consisting of a current state, an inciting incident, transformation and return. The current state describes the experience the user is in, introducing the user's values and concerns and the conditions that allow a relatable problem to emerge. The inciting incident describes the event where a problem emerges for the user. Following this, a proposition for a solution is introduced. The solution is described as transformation, showing the inciting incident being solved for the user. This is where the hero encounters or makes use of the envisioned product or service. The final component brings the story back to the starting point, with their initial challenge remedied.

Once the hero story is developed, it can be tested with prospective users or other stakeholders. After reading the story to participants, while recording their comments, open-ended questions should be asked to probe why participants are reacting in a certain way to the hero story.

The method is particularly valuable for the design of services that don't involve any physical or digital products that can be prototyped and tested. It can be used as a collaborative ideation method as well as for the evaluation of design concepts.

EXERCISE

YOU WILL NEED
Pen, paper, 2-4 people

In this exercise, you will develop a hero story, and use it to explore your design problem and generate feedback. Focus on your own design problem, or choose a design brief (p.138), and use the template on the companion website.

1 **Choose your hero**, the user that this story will focus on. Who are they? Describe their character and state before embarking on their journey. What are their values and concerns? This should be based on user research.
E.g. captured in a persona (p.100) or inspired by a specific observation.
[10 minutes]

2 **Start at the beginning**. What is your hero doing just before their problem emerges? Pair visuals with simple phrases to set the scene. Use sketches or photos to accompany each stage of the story.
[10 minutes]

3 **Write the inciting incident**. Something changes – the hero encounters an unexpected difficulty or learns something new that affects their view of the world. An action is required.
[10 minutes]

4 **Introduce the transformative solution**, describing how your hero takes action to solve the problem and what products or services support him. The solution does not have to be feasible; this is primarily used to test your understanding of the user's problem as represented by your hero.
[10 minutes]

5 **Conclude with the hero's return**. The final part of the story will wrap up the narrative, with the hero returning to an improved version of the situation where they started.
[10 minutes]

6 **Test the hero story with participants** by asking them to read the story out loud. While reading they can ask questions or express their opinions. Follow up with an open-ended interview and record their feedback.
[10 minutes per participant]

7 **Iterate and test again**. Based on your participants' feedback, what changes do you need to make? If your story didn't resonate with your participants go back and iterate the story based on their feedback, or change the envisioned solution.

Design. Think. Make. Break. Repeat.

Heuristic Evaluation

Testing your solution with domain experts

ACADEMIC RESOURCES:

Nielsen, J., & Molich, R. (1990, March). Heuristic evaluation of user interfaces. In Proceedings of the SIGCHI conference on Human factors in computing systems (pp. 249-256). ACM.

Nielsen, J. (1992, June). Finding usability problems through heuristic evaluation. In Proceedings of the SIGCHI conference on Human factors in computing systems (pp. 373-380). ACM.

Nielsen, J. (1994). Heuristic evaluation. In Nielsen, J., and Mack, R.L. (Eds.) Usability Inspection Methods. New York, NY: John Wiley & Sons.

Heuristic evaluation is a useful method for collecting feedback on early designs, within a short time span and at low cost. In contrast to other usability evaluation methods that involve end users, the heuristic evaluation method collects feedback from experts. Using their domain knowledge, experts establish whether the designed solution complies with certain usability principles – also referred to as heuristics.

Heuristic evaluation is performed using a checklist of elements. A popular reference checklist is the list of "10 Heuristics for User Interface Design" (Nielsen & Molich, 1990; Nielsen, 1994). The list includes visibility of system status, match between system and the real world, user control, consistency, error prevention, recognition rather than recall, flexibility of use, aesthetic, recovery from errors, and help option.

The domain expert interacts with the system, looking out for situations where the design doesn't conform with the heuristics. It is advised for the expert to interact with the product more than once to ensure sufficient familiarity prior to evaluation. The more experts are involved, the more thorough the results. According to studies, a single expert will find about 35 percent of all usability problems, while five experts will find about 75 percent of all problems (Nielsen & Molich, 1990). The number of issues identified also depends on the level of expertise of the evaluators (Nielsen, 1992).

The outcome of a heuristic evaluation is a report comprising a list of usability problems associated with system elements and categorised according to usability principles. The usefulness of the method depends heavily on the knowledge of the experts and their familiarity with the heuristics.

EXERCISE

YOU WILL NEED
Pen, paper

In this exercise, you will assume the role of an expert evaluator. You will evaluate a website of your choice, based on '10 Heuristics for User Interface Design'. Focus on your own design problem, or follow the 'Supermarket of the Future' brief (p.143), and use the provided template (p.177).

1 **Spend some time browsing the website**, making yourself familiar with the menu and different functions. Give yourself at least three different tasks to carry out:
E.g. order a loaf of bread.
E.g. find the cheapest brand of butter.
E.g. review the contents of your cart.
[15 minutes]

2 **Use the heuristic evaluation template** to evaluate the website. Go through the same tasks you just performed and make connections to each heuristic principle, noting any issues or problems:
E.g. if you don't receive any feedback when you click on 'GO', record this comment under Visibility of System Status.
E.g. if important buttons change position from page to page, record this comment under consistency.
[50 minutes]

3 **Go through the checklist** and make sure you have considered all heuristic elements. Ideally, you should consider each principle for every possible task. However, for the purpose of this exercise, you can focus on reviewing the three tasks that you have already made yourself familiar with.
[5–10 minutes]

4 **Review all usability problems** and their link to heuristic elements. Are there usability problems that do not fit into any of the ten given categories? Add these to the bottom of the list and try to define what is confusing about them, creating your own heuristic element.
[5–10 minutes]

5 **Add a severity rating** for each of the identified usability problems. Use a scale from 0 to 4:
0 = no usability problem
1 = cosmetic issue
2 = minor issue
3 = major issue
4 = usability catastrophe
[10–15 minutes]

6 As a follow-up activity, finalise the exercise by identifying ways in which you can **fix the usability problems**. Use brainstorming techniques such as brainwriting 6-3-5 (p.28). Record your ideas in the form of wireframes (p.136).

Design. Think. Make. Break. Repeat.

Interaction Relabelling

Shifting focus from functionality to interaction possibilities

ACADEMIC RESOURCES:

Djajadiningrat, J.P., Gaver, W.W. and Fres, J.W., (2000), August. Interaction relabelling and extreme characters: methods for exploring aesthetic interactions. In Proceedings of the 3rd conference on Designing interactive systems (pp. 66-71). ACM.

In some design situations, it can be a useful exercise to explore new ideas through lateral thinking as an alternative to making research-driven design decisions. The playful exploration of ideas can deliver insights that offer different qualities and perspectives when compared to those derived from analysing real data from real participants and contexts.

As an ideation method, interaction relabelling encourages the exploration of design ideas by substituting the to-be-designed product or service with another existing object. Interactions with the existing object are mapped to interactions with the envisioned product. The existing object does not have to closely resemble the to-be-designed artefact. It can also be useful to explore different existing objects of varying complexity. Mechanical products work well as they encourage thinking beyond the digital aspects of products or services.

For example, a toy revolver can serve as an object to examine interactions with a calendar (Djajadiningrat et al., 2000). Possible mappings include associating bullets with appointments, allowing for further associations such as removing the bullets to cancel all appointments or firing bullets at a wall to display the details of an appointment.

Ideally, interaction relabelling is performed in a group, as this encourages a playful competitive exploration of mapping interactions, enabling ideas to build on each other. Ideas generated through this method can then be introduced back into the design process and evaluated in terms of their value and meaningfulness within the context of the design problem. As such, interaction relabelling can provide an additional source of ideas and inspiration that complement other more structured methods.

EXERCISE

YOU WILL NEED
3–4 people,
mechanical object
(e.g. stapler, glasses,
umbrella), pen, paper

In this exercise, you will use interaction relabelling to brainstorm ideas for a new social networking application or a design brief you are working on. Use the glasses provided in the resources on the companion website or use another mechanical object to explore interactions.

1. **Warm-up by discussing** your design brief with your group.
E.g. what social networks do people in your group use?
E.g. what is the main thing that you use them for?
E.g. what are the activities that you perform during a typical session?
[5 minutes]

2. **Identify features** that are essential for interacting with this product. Try to make a list of at least ten different features. Take notes on paper or in your sketchbook.
E.g. ability to add and remove friends
[5 minutes]

3. **Imagine your chosen mechanical object was your way of interacting** with the social network. Pass the object around in your group and play out possible interactions. **Map the different tasks** you need to perform to specific features of the object.
E.g. putting on the glasses and looking at a person might add them as a friend.
[10 minutes]

4. **Continue discussing possible interactions** with your group, until you have exhausted all your ideas. **Identify interaction possibilities** that could be useful to take back into the design process, and make a note of these.
[10 minutes]

5. **Repeat the exercise** with a different object as the stimulus.

Design. Think. Make. Break. Repeat.

Interviews

Only by asking good questions, will you get good answers

ACADEMIC RESOURCES:

Doody, O., & Noonan, M. (2013). Preparing and conducting interviews to collect data. Nurse researcher, 20(5), 28-32.

Jacob, S. A., & Furgerson, S. P. (2012). Writing Interview Protocols and Conducting Interviews: Tips for Students New to the Field of Qualitative Research. The Qualitative Report, 17(42), 1-10.

Interviews are one of the most flexible research tools available to designers. They can be used in many phases of the design process to gather information from experts, users and other stakeholders. The aim of an interview may be to discern background information about a problem area, gauge users' opinions about concepts, or collect detailed feedback about a new prototype. With well-crafted interviews we can gain deep insight into the experiences of users and develop empathy with the people we are designing for. Interviews are particularly useful for gathering concrete information about existing experiences rather than speculation about future products or situations.

There are three common types of interviews. In structured interviews, the script is fixed in advance and closely followed during the interview. Unstructured interviews use mostly open-ended questions and questions emerge through the conversation. Semi-structured interviews use a combination of fixed script questions and open-ended questions. The type chosen in a design situation depends on the intent, for example, whether it is important to collect answers on specific questions or to broadly investigate a topic area. A less structured interview has the advantage of allowing the interviewer to 'probe' participants for additional information and to follow interesting tangents in the conversation.

Semi-structured or structured interviews typically take about an hour per participant; a minimum of three to eight participants is recommended for a small study. It is helpful to pilot the interview by trying it out with one person before committing to the script. It is also important to choose participants carefully to ensure they represent the target audience – close friends or colleagues will not be suitable if they are not users of the product or service being designed. Relevant participants will give relevant insights.

EXERCISE

YOU WILL NEED
A partner, pen, paper, audio recorder (recommended)

In this exercise, you will collect real data by conducting a semi-structured interview with another person about their experiences with a product or service. You will follow a basic script but are expected to pursue interesting leads. Use the provided template from the companion website to take notes.

1 **Choose a topic** to interview your partner about, for example, going to the cinema or taking public transport. You can also choose one of the design briefs as your focus topic (p.138).

2 Start by **warming up your participants** with simple questions. One way is to gather demographic data relevant to your topic.
E.g. name, age, gender, cultural background
[2–5 minutes]

3 **Use the following open-ended questions** to conduct your interview. These can be tailored to your topic:
- Tell me about the last time you went to …
- What were the best parts of the experience?
- What caused you frustration or prevented you from doing something?
- What motivates you to use/go to …
- What would you like to change about the experience?

[10–60 minutes]

4 **Prompt your interviewee** to provide more detail when they hit on an interesting topic. Try using the laddering method (p.82) to discover their deeper motivations and underlying reasons.

5 **Take notes as you go.** Practice recording keywords and pertinent details, without disturbing the flow of your interview by pausing for too long. In real interviews, the interviewer will often use a recording device to assist, in which case it is necessary to seek permission from the interviewee to record their responses.

6 You can **use the data gathered** from interviews in many ways, such as creating an affinity diagram (p.22), developing personas (p.100), or reflecting on what this information means for your design.

Design. Think. Make. Break. Repeat.   79

KJ Brainstorming

Collaboratively connecting and prioritising ideas

ACADEMIC RESOURCES:

Kawakita, J. (1991). The original KJ method. Tokyo: Kawakita Research Institute.

Spool, J. M. (2004). The KJ-technique: A group process for establishing priorities. User interface engineering.

When working in teams, it can be challenging to reach consensus about what should be prioritised in the design process. Having consensus is essential to allow teams to manage resources effectively by focusing on the most important issues or ideas. The "KJ Brainstorming" method provides a structured approach for identifying these top priorities.

Named after its inventor, Japanese anthropologist Jiro Kawakita, the KJ method provides a process for organising and prioritising a large number of data items that need to be addressed in a design solution – represented either in the form of issues or ideas. Items are grouped into core themes represented through an affinity diagram. Unlike the affinity diagramming method, which is used to analyse research data and generate ideas (p.22), the KJ method focuses on the collaborative identification of those core themes. Once the grouping of all items has been completed, participants collaboratively prioritise the three most important themes, which can represent either issues or ideas.

The KJ method gives clear instructions as to when the team should or should not discuss the items and groups. By having discussion-free time, the team can focus on the generation of themes without getting stuck on one issue or engaging in a conflict. The decision about the top themes is made collectively through voting.

While the KJ method is traditionally used for data organisation, it is a useful method for brainstorming new ideas when resolving a design problem. This is particularly useful when working with large teams and in situations where team members are already familiarised with the design problem.

EXERCISE

YOU WILL NEED
4+ people, pen, paper, post-it notes (3 colours), a wall, highlighter

In this exercise, you will use the KJ Brainstorming method to identify and prioritise themes in user research data. Use the resources on the companion website if you don't have your own data. Alternatively, you can also complete the exercise with ideas generated through the brainwriting 6-3-5 method (p.28).

1 **Read the interview transcripts** individually. Highlight statements in the text that refer to user's interests, needs, issues, behaviours, etc.
[10-15 minutes]

2 As a team, **choose a question** around the design issue as the focus of the affinity diagram.
E.g. 'What features in Airbnb or a calendar app would be important for business travellers?'.
Write the question on a piece of paper and hang it on the wall.
[2 minutes]

3 **Generate ideas individually** in response to the focus question, keeping in mind the user insights you read about. Write each response item on a Post-it note and stick it on the wall.
[10 minutes]

4 **Read all the notes** on the wall and add new items if desired. There should be no discussion within the group about the content of the notes at this point.
[5-10 minutes]

5 **Find items that belong together** and move them into new clusters on the wall. You may move items to groups created by others. Discussions should be avoided at this point. This phase continues until every note has been added to a cluster.
[5 minutes]

6 **Choose labels for each group** and record them on Post-it notes with different colours. You may discuss the labels with your team. Split, merge or create subgroups if necessary. Review the items in each group, and move them to different clusters if they don't belong anymore.
[5-10 minutes]

7 **Rank the three most important groups** that best match the focus question. Each team member votes for up to three groups by drawing an asterisk on groups that they believe should be prioritised. Discuss the emerging themes as a group and review ways they could be incorporated into designs.
[5 minutes]

Design. Think. Make. Break. Repeat.

Laddering

Finding out what really matters

ACADEMIC RESOURCES:

Jordan, P. (2000). Designing Pleasurable Products: An Introduction to the New Human Factors (pp. 121-181). London: Taylor & Francis.

Reynolds, T. J., & Gutman, J. (1986). Developing a Complete Understanding of the Consumer: Laddering Theory, Method, Analysis, and Interpretation. Oslo, Norway: Institute for Consumer Research.

The laddering method is an interview technique used to discover the underlying reasons for people's views about a product or service. Most people can easily articulate what they do or don't like about a product or service regarding its basic attributes. However, it is more difficult to identify the reasons; how the direct consequences of these attributes affect them and how this relates to their values. This is what the framework of the laddering method offers.

A laddered interview follows the same basic principles as a regular interview (p.78) but applies predefined levels of abstraction when it comes to asking the questions and coding the resulting data. The levels are basic attributes (A), consequences (C), and values (V). Laddering allows us to probe deeper into users' feelings while moving up the levels of abstraction, from attributes to consequences to personal values.

When conducting the laddered interview, the interviewer responds to the user's answers by asking why specific attributes are important to them. The discussion can be directed to the abstraction levels of underlying concerns. Following on from the interview, the next step is to transcribe users' responses and to code them according to the three levels of abstraction (A, C and V).

Laddering is useful for identifying the links between attributes, consequences and values and can add a unique perspective to design research. By simultaneously considering qualitative and quantitative aspects, laddering can add a unique perspective to design research. It allows the links between product attributes and perspectives, such as perceived benefits, values and attitudes, and can explain why certain attributes are more important than others.

EXERCISE

YOU WILL NEED
A partner, pen, paper

In this exercise, you will collect data by conducting a laddered interview with another person about a product or service. Focus on your own design problem, or choose a design brief (p.138).

1. **Choose a product or service** related to your chosen problem area as the focus for your laddered interview. Introduce this topic to your participant.
E.g. a taxi service (for the 'Autonomous Vehicles' brief, p.140)
[5 minutes]

2. **Write a list of open-ended questions** to use during your semi-structured interview. For suggestions of questions, refer to the interview method (p.78).
[10 minutes]

3. **Start the interview and listen carefully** to your participant's answers so that you can ask strategic questions. The initial reason that your participant gives for their feelings often relates to basic **attributes** of the product or service.
E.g. participant: 'I don't like using taxi services because I have to wait.'

4. **Ladder the interview** every time you hear a reason by asking: "Why is (insert reason here) important to you?" This will help you find the underlying **consequences** that have led to your participant forming this view.
E.g. interviewer: 'Why is not waiting important to you?'
Participant: 'Because I don't want to plan ahead, I'd just like to go when I want.'

5. **Keep asking this question** until the user runs out of answers. This will help you to **reach the underlying values** of the participant.
E.g. interviewer: 'Why don't you want to plan ahead?'
Participant: 'Because I want the freedom to be impulsive.'

6. **Continue steps three to six** throughout the entire semi-structured interview, until you run out of prepared questions and the conversation finishes naturally.
[60 minutes]

7. **Review the verbatim notes** of the interview, **identify and code** them into:
(A) Attributes: concrete and abstract
(C) Consequences: functional and psychological
(V) Values: instrumental and terminal.

8. **Construct concept maps** to summarise your data, by drawing links between coded words.
E.g. freedom --> not planning --> not waiting

Design. Think. Make. Break. Repeat.

Low-fidelity Prototyping

Creating tangible representations of ideas

ACADEMIC RESOURCES:

Sefelin, R., Tscheligi, M., & Giller, V. (2003). Paper prototyping-what is it good for?: a comparison of paper-and computer-based low-fidelity prototyping. In CHI'03 extended abstracts on Human factors in computing systems (pp. 778-779). ACM.

Svanaes, D., & Seland, G. (2004). Putting the users center stage: role playing and low-fi prototyping enable end users to design mobile systems. In Proceedings of the SIGCHI conference on Human factors in computing systems (pp. 479-486). ACM.

Warfel, T. Z. (2009). Prototyping: a practitioner's guide. New York, NY: Rosenfeld Media.

A prototype is a representation of an envisioned product. Low-fidelity prototyping allows the quick exploration of ideas early in a design process. They can be used to reflect on a design, to discuss design solutions within a team and to get feedback from prospective users through usability testing (p.126).

A prototype should represent the actual scale of the final product. This allows people to experience what the interaction with the final product would feel like, whether it is easy to use and whether there are aspects that need to be improved. They are typically hand-sketched, use various materials, and do not represent the final visual design. Jeff Hawkins, the founder of Palm Computing, carried around a block of wood the size of a pocket computer, which later led to the invention of the hugely successful PalmPilot. To determine what features to include and how to design them, he would take it to meetings and use a pen to interact with the prototype as if he was setting up calendar appointments.

Low-fidelity prototypes are different to mock-ups (p.90) as they do not represent the final visual design. They are also different from wireframes (p.136) as they use tangible materials to explore how people could interact with the final product or service. As such, low-fidelity prototypes often take the form of a 'horizontal' prototype, which is a prototype that represents the surface of a product interface, but not any of the underlying technology. In comparison, 'vertical' prototypes only represent a small aspect of the product interface (for example, only the login page) but are fully implemented and functional – which can be useful to test technical aspects, such as loading times.

EXERCISE

YOU WILL NEED
Pen, paper, tracing paper, scissors, sticky tape, Blu-tack

In this exercise, you will design a low-fidelity prototype that addresses a specific brief. Focus on your own design problem, or choose a design brief (p.138). If you have findings from user research (e.g. interviews, p.78), be sure to consider these in your design.

1 **Sketch** the parts of your product that you want to be **represented in the prototype**. Choose a specific task and sketch all the steps required to fulfil this task.
E.g. adding a person to your friends in a social network app.
[15 minutes]

2 **Create wireframes** (p.136) of your product using pen and paper. These are an iteration of the sketches produced in the first step and show the key elements on each screen.
[30 minutes]

3 **Create a physical prototype** based on the previous step using any suitable materials.
E.g. for the social network app example, use the smartphone drawing templates provided (p.195) to create a paper version of each screen.
Add dynamic features to allow for user interaction. This can be achieved either by swapping out a screen or by adding layers in the form of Post-it notes.
E.g. in a social network app, pressing the 'Find' button will bring up a new screen that lists people matching the name that was 'typed' into the search field.
[30 minutes]

4 **Add real content** by writing it directly into your prototype or on additional materials that can be overlaid, such as Post-it notes.
E.g. in a screen showing search results in a social network app, add names that are made up but sound real. This will help to give the prototype a more realistic feel
[15 minutes]

5 You can also **prototype other physical aspects.** For example, you could create a physical representation of a virtual reality headset with minimal materials
E.g. using a cardboard VR headset

Design. Think. Make. Break. Repeat.

Mapping Space

Capturing bodily movement through space and time

ACADEMIC RESOURCES:

Lynch, K. (1960). The image of the city (Vol. 11). Cambridge, MA: MIT Press.

Paay, J., Kjeldskov, J., Howard, S., & Dave, B. (2009). Out on the town: A socio-physical approach to the design of a context-aware urban guide. ACM Transactions on Computer-Human Interaction (TOCHI), 16(2), 7.

For more than a century, artists, architects and photographers have been concerned with the pursuit of recording the bodily movements of humans through space and time. Before video cameras they had to make do with static means. This fascination with how to capture motion over time led to the Italian art movements of dynamism and futurism, which upheld the ideals of capturing energy on the canvas or in sculptural form.

However, the ability to map movements in a static way has practical applications beyond artistic expression. As a research technique, it helps us to record information about the qualities of a space, how it is utilised, what the common pathways and flows are, and what volume the human body occupies within that space. In a single static snapshot, we can gain a clear impression of how a space is used over time by multiple people and by individual users. These considerations are useful when designing an object or system with spatial dimensions.

There are many dimensions of space that we can map, such as the boundaries of that space, the choreographies of people moving through it and its proportions. Also of interest may be the infrastructure and connections between areas, trajectories and speeds of objects, as well as the insides of voids and outsides of volumes. This is typically done through different kinds of linework (colour and thickness) on paper, in a process where we build up the information through translucent layers. The method can involve any form of representation that helps to document these dimensions.

EXERCISE

YOU WILL NEED
A partner, marker, coloured pens, paper, tracing paper, camera

In this exercise, you will create a static map of a physical space that embodies multiple dimensions of how that space is used. Increase your perceptual and analytical skills by trying out different methods for mapping space during the exercise.

1. **Go to the space** you will be mapping, spend time immersing yourself in it. Measure it with your body by walking through it and counting your steps. Discuss the space and its qualities with your partner. What are the textures, materials, lighting? Are there plants, animals, people?
[5 minutes]

2. **Map the boundaries of the space with a thick black line** including outlines of walls/boundaries, furniture or functional elements. Make several copies of this map, so you can create multiple versions with different information.
[10 minutes]

3. On a layer of tracing paper **map the use of the space with coloured lines**. Trace different people's trajectories. Include a chronological aspect in this; who uses the space and when? Signify different uses or users with different colours.
[25 minutes]

4. **Take detailed photos** of people using the space. Describe and record what activities are happening. Take multiple photos from the same perspective.
E.g. at 5:13 pm, a young man arrives and takes out a sandwich, eats it and leaves at 5:23 pm.
[10 minutes]

5. **Exchange your map** with your partner and map additional information on theirs on another layer of tracing paper. Try to add a new perspective or additional details. Observe the different perspectives they have compared to yours.
[10 minutes]

6. **Map the movements of bodies** in the space by:
   - Taking multiple photos of a single person's movement through the space.
   - Taking multiple photos of a person's gestures in relation to the space.
   - Taking multiple photos of two or more people interacting.
[30 minutes]

7. **Complete the documentation** of your space by adding your photos to your map, at the locations where the observations were made. This could be done on paper, or using computer software. You could also consider superimposing your photos over one another, to capture people's dynamic movements in a static way.

Design. Think. Make. Break. Repeat.

Mindmapping

(Who, What, When, Where, Why, How)

Making a snapshot of connections and relationships

ACADEMIC RESOURCES:

Buzan, T., & Buzan, B. (1996). The Mind Map Book: How to Use Radiant Thinking to Maximize Your Brain's Untapped Potential, New York: NY: Plume.

Kokotovich, V. (2008). Problem analysis and thinking tools: an empirical study of non-hierarchical mind mapping. Design Studies, 29(1), 49-69.

Mindmapping is a notetaking and notemaking technique that developed in the 1970s, emerging from cognitive research about how our brains interpret and store information. In mindmapping, we record information in a way that utilises physical and spatial relations to better facilitate long-term retention. The technique has several applications in design, as we are often dealing with large amounts of information which makes it difficult to retain an overview.

While there are a number of 'rules' for creating mind maps, the method has existed sufficiently long for others to find additional ways of leveraging it. However, the initial key rules as expounded by Buzan and Buzan (1996) are still relevant, as it is precisely this way of presenting and structuring the information that helps us to retain it. Their key rules are as follows:

- The central concept is recorded in the middle of the page, as an anchor for branches.
- Each new addition is attached to the anchor or an existing branch; there should be no free-floating elements.
- Colours and images are used to increase memorability.
- Keywords are used on each branch instead of full sentences.
- Text is recorded following a consistent direction: no vertical or upside-down text.

A version of mindmapping that is particularly useful in design is when information is structured under the key branches of 'who', 'what', 'when', 'why', 'where' and 'how' – referred to as the WWWWWH framework. By adding all the information that we know about a problem area, we can very quickly build up a map of the context in which a new product or service is to be designed – and start generating ideas using these triggers. This is a great way to get existing information out of your brain and onto the page.

EXERCISE

YOU WILL NEED
Pen, paper, coloured pencils

In this exercise, you will create a mind map step-by-step, using the WWWWWH framework, in order to build up a quick overview of your problem area. Your finished mind map will serve as a reference for ideation and future design work.

1 **Write your central theme or topic** in the middle of a piece of paper.

Write down evenly spaced branches radiating from the central theme with the keywords 'who', 'what', 'where', 'when', 'why', 'how'.

2 **Add all the information** about the problem area to these categories. In order to fill each branch, **ask yourself questions** that start with that specific word:

- **Who** is using the product or service? Write down all the possible users or stakeholders.
- **What** format or features does the product or service have? What is it used for?
- **Where** are they using it? Inside, outside, in a specific location? On their body?
- **When** are they using it? This could be a time of the day, or during an activity.
- **Why** are they using it? High level and functional goals are equally useful here.
- **How** do they use it? This sometimes already includes modalities or methods of interaction like buttons or handles, and may sometimes include some initial ideas too.

[15 minutes]

3 Each question can have multiple answers. **Record all of them down** in the mind map. Don't be concerned if answers in the same category don't make sense together or are not the same sort of answers. Remember to use keywords to record your answers, not full sentences.

4 Use your mind map as **a tool for ideation**, for example, by using forced associations (p.66). The mind map can also be used to reflect on whether your design concepts are tackling a sufficient number of aspects of the problem area. Use it as an ongoing reference for your design project.

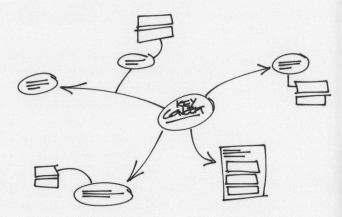

Design. Think. Make. Break. Repeat.

Mock-ups

Creating visual models of envisioned products

ACADEMIC RESOURCES:

Gerber, E., & Carroll, M. (2012). The psychological experience of prototyping. Design studies, 33(1), 64-84.

Ohshima, T., Kuroki, T., Yamamoto, H., & Tamura, H. (2003). A mixed reality system with visual and tangible interaction capability- application to evaluating automobile interior design. In Proceedings of the 2nd IEEE/ACM International Symposium on Mixed and Augmented Reality (p. 284). IEEE Computer Society.

Moc-kups are scale or full size models of products and their features. They have been used widely in the industrial design and manufacturing process to visualise the detailed representation of an envisioned product early on in a design process. They are also useful for testing ideas and getting buy-in from external stakeholders, before spending time on building functional prototypes.

Mock-ups focus on visual aspects, such as style and colour, whereas wireframes (p.136) focus on structure and functionality. When designing physical products, mock-ups allow for the exploration of specific form factors, such as how people would hold the designed product. When designing digital solutions, they are used for creating high-fidelity representations of the visual design, which include actual content, font type, colour scheme, and so on. The design of a mock-up should resemble as closely as possible the final visual representation of the product. The process of creating mock-ups depends on the nature of the product. Digital-only solutions can be produced in vector-based graphics editing tools or with specialised mock-up software tools that offer predefined user interface elements. Physical products can be mocked-up using 3D modelling software and rapid prototyping techniques, such as 3D printing and laser cutting, and subsequently applying colour and texture features. It is also possible to use digital projection mapping for adding colour and visual textures onto physical mock-ups. This enables designers to try out different colours or visual designs at the click of a button.

Mock-ups don't allow for user interaction as they aren't functional. Nonetheless, mock-ups can be used for preliminary user testing of certain features of a product and even to simulate user interaction and flow. When testing digital mock-ups with participants, it can be useful to use print-outs of the mock-up screens.

EXERCISE

YOU WILL NEED
Sketch (sketchapp.com) or an alternative vector-based graphics tool

In this exercise, you will create a series of mock-up screens for a digital solution. Start with existing wireframes, either from your own design problem, or follow the 'Autonomous Vehicles' brief (p.140). The wireframes provide the structure, functions and placeholder content and can be found in the resources on the companion website.

1 **Select three wireframes** that depict a use scenario for a digital solution, either from one of your projects or the resources section on the companion website.

2 **Establish style and colour.** Make sure that these match the nature of the product.
E.g. for a banking app, choose a corporate colour scheme.
E.g. for a social networking app, choose an informal colour scheme.
If you introduce specific graphical details, establish rules for when and where these will be used.
E.g. rounded corners on all buttons
[15 minutes]

3 **Decide on one or two font types and sizes**, and use those consistently throughout your mockup. The same style should be used for the- same purposes on each screen.
E.g. bold 12pt Helvetica for menu titles, 10pt regular Helvetica for body text
[5 minutes]

4 **Establish the content.** You can make up the content, both text and images, but you need to ensure that it looks as realistic as possible. For example, use photos of real people from a free stock image website, not cartoons or silhouette representations.
[20 minutes]

5 **Apply style, colour and content to all three wireframes**, creating your own mock-ups in Sketch or your graphics tool of choice. Start with one screen and iterate on style and colour until you are happy with the result. Only then should you start working on the other two screens.
[80 minutes]

6 **Try out your mock-ups.** They can be used to present your design ideas in a design critique session (p.52) or for usability testing (p.126). If you have trouble making choices about specific features, you can also make two alternative versions and perform an A/B test (p.20) to see which one produces a better result.

Mood Boards

Collecting visual inspiration

ACADEMIC RESOURCES:

Chang, H. M., Díaz, M., Català, A., Chen, W., & Rauterberg, M. (2014, June). Mood boards as a universal tool for investigating emotional experience. In International Conference of Design, User Experience, and Usability (pp. 220-231). Springer, Cham.

Lucero, A. (2012, June). Framing, aligning, paradoxing, abstracting, and directing: how design mood boards work. In Proceedings of the Designing Interactive Systems Conference (pp. 438-447). ACM.

McDonagh, D., & Storer, I. (2004). Mood boards as a design catalyst and resource: Researching an under-researched area. The Design Journal, 7(3), 16-31.

Mood boards are used to visually represent design ideas, concepts and values in an open-ended and evocative way. There are no specific formulas for designing mood boards. However, they should be crafted to foster discussion and to facilitate the generation of ideas. The final outcome looks similar to a collage, comprising images from magazines, photographs and other visual elements, such as coloured paper, textures, fabrics and any other object that could be easily attached to a blank surface or board. It is also possible to create digital mood boards, which may include video snippets, website captures, various images, and so on.

Crafting an expressive and communicative mood board takes time and reflection. Before starting with the collection of images and objects that will compose your mood board, it is a good idea to have a team discussion about the qualities and values your product or service should represent. The use of additional tools such as brainstorming or mind maps can help with the initial articulation of these concepts. Once you have reached an agreement about these general points, you can proceed to search for images. New ideas may emerge while collecting images and composing a mood board.

Mood boards show ideas in evocative and suggestive ways to encourage discussion and idea generation. When used to define a specific visual style, these tools are helpful to offer a reference point to stakeholders and designers. To communicate the meaning behind your creation you should rely on your spoken, rather than written, expression, as the use of specific annotations and written explanations can limit the openness offered by the tool. A good balance between abstraction and concrete inspirations is desirable for the final composition (Lucero, 2012).

EXERCISE

YOU WILL NEED
2 people, old magazines/newspapers, coloured paper, craft materials with different textures, A3 cardboard, A3 paper, scissors, glue, pen, notebook

In this exercise, you will craft a mood board to provide visual inspiration for a design project. Focus on your own design problem, or choose a design brief (p.138).

1 **Brainstorm possible design ideas** with your team. Discuss the kind of experience you want to achieve. Document the values and core qualities of this experience as **keywords** in your notebook. Select one idea to focus on, and think of a catchy title to represent it:
E.g. 'the office on wheels': values/qualities = convenience, familiarity, comfort
E.g. 'the 30-second shopping trip': values/qualities = speed, efficiency, dynamism
E.g. 'the museum for babies': values/qualities = curiosity, encouragement, exploration
[15 minutes]

2 **Collect and cut out images** from magazines. Allow yourself to be inspired; don't go hunting for a specific image. Think of images as ways to represent metaphors, experiences, visual aspects and material qualities of your idea. Discuss the relevance of each image with your team.
[100 minutes]

3 **Lay out your images on the A3 cardboard**. Once you are pleased with the composition, stick them down with glue. You can also make more than one mood board – documenting different aspects of your design concept.
E.g. visual aspects: what does it look like?
E.g. experiential aspects: what is it like to engage with it?
[15 minutes]

4 **Present your mood board to another team**:
- For the team **presenting the idea**: Mention your idea and the keywords that inspired your final piece, as well as any metaphors or choices made. Take note of the feedback you receive.
- For the team **giving feedback at the end of the presentation**: Put yourself in the shoes of a group of stakeholders. Ask questions about the mood board. Highlight positive aspects of it. Suggest changes and possible refinements.

Switch roles after the discussion.
[5 minutes per team]

Design. Think. Make. Break. Repeat.

Online Ethnography

Gathering insights from online communities

ACADEMIC RESOURCES:

Campbell, E., & Lassiter, L. E. (2014). Doing ethnography today: Theories, methods, exercises. New York, NY: John Wiley & Sons.

Postill, J., & Pink, S. (2012). Social media ethnography: The digital researcher in a messy web. Media International Australia, 145(1), 123-134.

Ethnographic research entails the use of qualitative methods for collecting data about a target audience. Traditional ethnography involves the researcher becoming part of a community and collecting field data through observations and interactions, overtly or covertly. The strength of conducting an ethnographic study lies in the richness of the data collected, which can deliver insights that would not be possible to gather through other methods alone, such as interviews (p.78).

Online ethnography translates the same principles to the study of people and their interactions online. It is therefore especially well-suited for studying user groups that are active online. Any online platform that allows people to post content and that is openly accessible can be used. Useful platforms include social networking sites (e.g. Twitter), online forums (e.g. TripAdvisor) or online news articles with comments sections. Similar to traditional ethnography, the researcher can either passively observe and collect data posted online or become an active member of the online community, post content and interact with other members. It is important to consider ethical and privacy implications, as the participants in an online ethnography are not always informed about their participation. Where possible, consent should be sought.

Text, images, videos and other content can be collected by either reading community posts online, or by accessing and aggregating content using a platform's application programming interface. Observations about people's behaviour, their interactions and their opinions can be recorded in a field diary. The data can then subsequently be summarised into main themes, for example, by using affinity diagramming (p.22). Potential research bias can be addressed by involving a partner in the interpretation of the findings.

EXERCISE

YOU WILL NEED
Internet access, pen

In this exercise, you will conduct online ethnography using a relevant online platform. Use the provided template (p.178) to collect your field notes. Focus on your own design problem, or follow the 'Designing Space Travel' brief (p.141).

1 **Choose an online platform** to focus on while you conduct your online ethnography. You may need to do some initial research to find out what platforms your prospective users are active on. Identify and familiarise yourself with this platform.
E.g. you can choose TripAdvisor for the 'Designing Space Travel 'brief.
[5-10 minutes]

2 **Decide on a specific set of criteria** to guide your ethnographic approach. For example, if you selected TripAdvisor, you might consider all reviews posted for a specific airline in the last two months. Or you might filter your results to focus on people living in Sydney, Australia, who have visited at least 100 cities.
[5-10 minutes]

3 **Read the entries** posted on the online platform and **make a note of your observations** in the template (p.178). This is the beginning of your own field diary. Your field notes will consist of a variety of different observations, such as:
- Records of what people say
- Attitudes of different users
- Common themes that emerge regularly in discussion
- Social interactions between people across posts

[30-60 minutes]

4 **Analyse your field notes** by creating an affinity diagram (p.22) or using the thematic analysis method (p.122). You may want to pursue particularly interesting insights by performing additional online ethnography, or by looking at people's behaviour on another relevant platform for comparison.

Design. Think. Make. Break. Repeat.

Perceptual Maps

Capturing the current market landscape

ACADEMIC RESOURCES:

Ferrell, O., & Hartline, M. (2013). Marketing Strategy, Text and Cases: Cengage Learning.

Gelici-Zeko, M. M., Lutters, D., Klooster, R. T., & Weijzen, P. L. G. (2013). Studying the influence of packaging design on consumer perceptions (of dairy products) using categorizing and perceptual mapping. Packaging Technology and Science, 26(4), 215-228. John Wiley & Sons.

Perceptual maps are graphical representations of how people subjectively perceive products or consumer brands available in the market. Products or brands are positioned or mapped according to attributes that are relevant for customers, such as quality or price. The nature of attributes is determined by the specific market that is being catered for. For instance, attributes such as comfort and safety might be fundamental for products relating to transport, whereas battery life, storage capacity and portability are likely more relevant for digital consumer products. Attributes can also refer to values that products or brands represent. These attributes are expressed as opposite adjectives such as 'family friendly' versus 'adult', 'simple' versus 'sophisticated', and so on.

Before building a perceptual map, it is important to identify the potential competitors. After this step, data is collected from potential customers about how they perceive the competitor brands or products. For instance, surveys including uneven Likert scales or semantic differential scales are useful to rate how attributes represented by each brand are perceived. After the data is tabulated and each brand is rated, groups of two relevant attributes are selected. Customers' perceptions are then mapped on a four-quadrant diagram illustrating the selected attributes, which are represented as polar opposites.

Perceptual maps are useful during the discovery phase to identify niches, trends and opportunities for design. It is important to take into account that brand perception varies over time. Therefore, perceptual maps should be considered as snapshots of the market landscape at the time when they were created.

EXERCISE

YOU WILL NEED
Pen, a partner, 10+ participants

In this exercise, you will collect data about user perceptions of different products or brands. You will then create a basic perceptual map with two key dimensions, using the provided template (p.179).

1 **Identify** market competitors related to your chosen design problem or use the sample competitors provided in the template (p.179).
[5 minutes]

2 **Generate a list of possible attributes of interest**. These attributes should be aligned with the type of market sector you are researching and relevant to your users. Fill in each semantic differential scale on the template by pairing up opposing attributes.
E.g. 'cheap' versus 'expensive'
[10 minutes]

3 **Collect data** using copies of your prepared template. Ask each of your ten participants to rate each product or brand according to their perception.
[20 minutes]

4 **Review the results, tabulating and calculating** the average score that each brand or product has obtained against each attribute.
[10 minutes]

5 **Use the grid at the bottom of the template** (p.179) to map the position of each brand. Choose a specific attribute for each axis. The centre of the grid (0,0) corresponds to a neutral point.
E.g. low-quality versus high-quality on the horizontal axis, and low price against high price on the vertical axis.
[5 minutes]

6 **Position each product or brand name on the grid**. You can include images if available or sketch the logos/products. You can repeat steps four to six using your remaining data, to compare different combinations of attributes.
[5 minutes]

7 **Identify which areas of the grid** are well-populated and which areas are empty. Discuss with your partner:
- Possible niches or design opportunities
- Any reasons why brands are not occupying empty spaces.
- How your map would change if your selected attributes were less polarised.

[5 minutes]

Design. Think. Make. Break. Repeat.

Persona-based Walkthroughs

Seeing designs through the eyes of users

ACADEMIC RESOURCES:

Nielsen, L. (2003). A Model for Personas and Scenarios Creation. In Proceedings of the Third Danish Human-Computer Interaction Research Symposium (pp. 71- 74). Roskilde, Denmark.

Pruitt, J., & Adlin, T. (2005). The persona lifecycle: keeping people in mind throughout product design. San Francisco, CA: Morgan Kaufmann Publishers.

'The Persona is static but becomes dynamic when inserted into the actions of the scenario. In the scenario the Persona will be in a context, in a specific situation, having a specific goal.' (Nielsen, 2003, p.72)

If personas (p.100) are characters that allow us to represent users, persona-based walkthroughs are the stage and story through which they are brought to life. A persona-based walkthrough gives these characters something to do. We enact a persona and 'walk' through specific tasks or scenarios, seeing the designs through their eyes. Thus, we become the minicab driver using their navigation app, the commuter on a bus playing a game on their smartphone or a pilot using specialist tools or software. We can evaluate designs from the perspective of those who will ultimately end up using them.

Walkthroughs originated as a way of evaluating the learnability of 'walk-up-and-use' systems like ATMs. As a tool, they stand-in for end users when involving real users may not be possible. Although they lend themselves to evaluation, persona-based walkthroughs can serve wider needs. For example, when designing a home security system, personas might be used to represent different homeowners and perhaps a burglar. A walkthrough may be used to evaluate existing security systems before embarking on the design phase. How easy is the alarm to disable if the owner accidentally sets it off? How hard is it to disable for the burglar? The type of scenario and how to apply personas to it depends on the stage of the design process and what questions need to be resolved.

EXERCISE

YOU WILL NEED
A partner, pen, paper

In this exercise, you will use a persona-based walkthrough to evaluate an existing product or service. You can choose a persona you have previously created for your own project or select the persona provided in the resources on the companion website.

1 **Choose a product or service** to evaluate. It should be:
- Accessible
- *E.g. a mobile phone app, website, self-service checkout.*
- Observable: visible, open to observation
- Realistic: the combination of persona and scenario needs to be realistic and relevant to the design goals. You would not test a first person shooter game with a persona who would never play video games.

2 **Assign roles** to yourself and your partner – **the actor and the evaluator**:
- Actor: the person who will act as the persona and do the walkthrough.
- Evaluator: the person who will observe the persona-based walkthrough.

3 The evaluator **selects a task for the persona to walk through**. The task should have a clear goal.
E.g. 'book movie tickets', 'play a game', 'search for a library book'
[5 minutes]

4 The actor **gets into character and performs the task** set by the evaluator. While undertaking the task, the actor may choose to think-aloud or voice the feelings and concerns of the persona.
[15 minutes]

5 The evaluator should **observe the persona walkthrough and take notes**. Identify sections of the task that are fluid, specific struggles, or times when the actor gets lost.

6 **Record and discuss** your findings together.
E.g. how did it feel to do the task as someone else? What did the evaluator notice? What issues caused frustration? What new design features could fix the issues that were observed?
[5 minutes]

7 **Record changes to the design** that you would like to make as a result of your findings from the persona-based walkthrough.
[5 minutes]

Personas

Modelling users through storytelling

ACADEMIC RESOURCES:

Cooper, A. (2004). The inmates are running the asylum:[Why high-tech products drive us crazy and how to restore the sanity]. Indianapolis, USA: Sams.

Courage, C., & Baxter, K. (2005). Understanding your users: A practical guide to user requirements methods, tools and techniques. San Francisco: Morgan Kaufmann Publishers

Goodwin, K. (2009). Chapter 11: Personas. In Designing for the Digital Age: How to create human-centred products and services (pp.230-297). Indianapolis, IN: Wiley Publishing.

Personas are fictional characters used to represent typical users, customers or other stakeholders. They are created from a synthesis of data from real people gathered through methods, such as interviews (p.78) and questionnaires (p.102). Personas can distil the information from our raw user data that is most pertinent to the design issues, avoiding idiosyncratic information that might be misleading.

Storytelling is used to thread together people's goals, motivations, attitudes and behaviours into a unified character. A persona called Alex, who is 23 years old, works in a bookshop, uses a MacBook Pro and is distrustful of buying clothes online, is more specific than an abstract and generic 'user'. This personal quality allows us to engage socially and emotionally with the needs of the user and to include their voice within all phases of the design process. It can also be valuable to have a non-persona included – a representation of a user who would not engage with the designed product or service.

Unlike real users, personas are more tolerant of rough early sketches and lengthy design meetings. They can be used to communicate user needs within the design team, troubleshoot design problems before reaching usability testing (p.126), and guard against basing design decisions on our own preferences and biases.

Personas are created by collecting data, creating variables, finding patterns in those variables and articulating them in the form of one or more visually engaging artefacts. A persona should communicate motivations, frustrations, attitudes, goals, behaviours and demographic information. Though the persona may contain fictional information, such as a name and a profile photo, it is not a piece of creative writing. A great story that fails to present the user's needs, fails in its purpose.

EXERCISE

YOU WILL NEED
Pen, paper, highlighters

In this exercise, you will create a persona using interview data. Identify variables in the data and look for patterns across the interviewees. Use the provided templates (pp.180–181). If you don't have your own data, use the resources on the companion website.

1 **Read through the interview data** and **identify important variables**. Add these to the variables sheet (p.180). Variables can be behavioural, attitudinal or demographic. Some example variables to look for include:
- Demographic data:
- *E.g. age, gender, dependents, marital status, job type, etc.*
- Behavioural variables:
- *E.g. frequency, cost sensitivity, tech savviness, etc.*
- Attitude variables:
- *E.g. likes/dislikes, trusts/mistrusts, emotional (de)attachment, etc.*

[20 minutes]

2 **Synthesise patterns in the data** by looking for patterns or clusters across all the variables in your interview data. This will manifest as similar behaviour across multiple people who share other characteristics.
E.g. interview data for a study app might reveal that students who plan their work in advance also tend to be older than those who cram at the last minute.

[15 minutes]

3 **Amalgamate patterns** in the data to form a single but coherent 'persona'. You may have found that 'last-minute crammers' are more likely to use Android and have a part-time job. Make the persona believable and engaging by using storytelling; provide a name and backstory. Use the persona creation template (p.181) to record your persona.

[20 minutes]

4 **Review and refine**. Review your persona by getting feedback from someone else and checking your interview data. Ask yourself the following questions:
- How realistic, convincing and coherent is the persona?
- Are the goals specific to the design problem?

[10 minutes]

5 **Use your persona**. Brainstorm ideas to help that persona achieve their goal, create a scenario (p.110) for the persona or use a persona-based walkthrough exercise (p.98) to evaluate a prototype or product.

Design. Think. Make. Break. Repeat.

Questionnaires
Gathering large amounts of user data

ACADEMIC RESOURCES:

Ballinger, C., & Davey, C. (1998). Designing a questionnaire: an overview. British Journal of Occupational Therapy, 61(12), 547-550.

Boynton, P. M., & Greenhalgh, T. (2004). Hands-on guide to questionnaire research selecting, designing, and developing your questionnaire. British Medical Journal 328, 1312-1315.

Colosi. L. (2006). Designing an effective questionnaire. Research brief available online at: http://www.human.cornell.edu/pam/outreach/parenting/parents/upload/Designing-20an-20Effective-20Questionnaire.pdf

Questionnaires collect information in written form and may be communicated by paper or digitally. They are a low-cost way for collecting large amounts of data without requiring trained facilitators or lab equipment. Questionnaires are commonly used during early user research and for collecting feedback about people's current experiences with existing products or services. A consent form can be included in the questionnaire to obtain permission for collecting data from respondents.

A good questionnaire can give designers insight into a person's self-reported behaviours, attitudes or perspectives. To achieve this, it is important to establish clear research questions in advance, and to specify the type of data that can answer those questions. It is also important to devise a plan for how to analyse the data collected through the questionnaire before distributing it.

Long questionnaires are less likely to be completed by participants and are likely to result in less valuable data. It is, therefore, important for every single question to be carefully selected, keeping in mind what it measures and how the knowledge created contributes to the design project. Any question that does not fulfil a purpose should be left out of the questionnaire. Ambiguous questions that do not reveal a clear link to the objective of the questionnaire should be avoided. Each question should contain only one topic and should not lead the respondent to answer in a certain way.

To ensure a questionnaire collects consistent and reliable responses a 'test-retest' can be administered, whereby a typical respondent is asked to complete the same questionnaire twice. Once the design of the questionnaire is complete, it is recommended to conduct a pilot test with a few respondents to test the clarity of the questionnaire.

EXERCISE

YOU WILL NEED
A partner, pen, paper, smartphone, computer (optional)

In this exercise, you will learn the basic steps of designing a questionnaire that collects feedback on the user experience of a product. If you don't have a product in mind, then choose an app on your smartphone for this exercise.

1. **Decide on the objective** for your questionnaire.
   E.g. to find out how often people use a specific smartphone application and in what circumstances.

2. **Consider what sort of participants** are required. Think about how you can distribute your questionnaire to these people. In some cases, paper copies will be best. In other cases, an online questionnaire may be more suitable.
   [5 minutes]

3. **Determine a set of relevant questions** to fulfil your objective. To ensure the right data is collected, it is a good idea to **start with standardised questions**. These can be found online on research databases such as allaboutux.org.
   E.g. to examine whether past experiences with similar products impact current experience, ask if the respondent has ever used a similar product.
   [20–30 minutes]

4. **Select your question types**, and place them in a meaningful order:
   - Demographic: relevant personal details that help to categorise data. *E.g. age, gender, occupation, etc.*
   - Open-ended: used for exploratory purposes when there is not a lot of data available on a topic that can be used to formulate the question; *E.g. what issues did you encounter when setting up the product?*
   - Closed: multiple choices that are used when there is existing knowledge about the topic. *E.g. have you used this product in the past three months? Yes/No.*
   - Likert scales: scales that measure in increments and generally range from 1 to 5, 7, or 9. *E.g. extent of disagreement-agreement, frequency (never-always), attitude (dislike-like), etc.*

   [20–30 minutes]

5. **Pilot test your questionnaire** to make sure the questions are understood the same way consistently by participants. Remove any remaining ambiguity by rephrasing where necessary.
   [15 minutes]

Design. Think. Make. Break. Repeat.   103

Reframing

Taking a *look from a different* perspective

ACADEMIC RESOURCES:

Dorst, K. (2015). Frame Innovation: Create New Thinking by Design. Cambridge, MA: MIT Press.

Seelig, T. L. (2012). inGenius: A Crash Course on Creativity. San Francisco, CA: Harpe One.

When we embark on a design project, we usually want to solve a specific challenge or come up with an innovative new idea within a problem area. It is easy but dangerous to get stuck on the first solution that comes to mind. Reframing is a way of learning to look at a problem from a different perspective in order to unlock a broader range of potential solutions.

When we define a problem, we automatically limit ourselves – just through our choice of words. A poorly formulated problem statement can give us tunnel vision by containing assumptions about what kind of solution is needed. Thus, we have already limited our creativity, shrunk our range of options, and denied ourselves the possibility of innovation – before we've even started.

For example, if the design brief is to 'design a new door', using the word door means that we are likely to stick with the kind of solutions that already exist; it will be a solid panel of material, probably attached to hinges or sliders. Perhaps our new door will have some innovative features, but most likely it will be in the realm of what already exists. But if we are asked to design 'a way to stop people entering a room' or 'a way to keep cold air out of a house', suddenly the range of possible solutions becomes broader. This is the basic principle of reframing.

Reframing techniques focus on changing the formulation of our initial problem statement, to make it broader and more abstract. When embarking on a design project, it is important to first question the problem before starting with finding a solution.

EXERCISE

YOU WILL NEED
Pen, paper

In this exercise, you will reframe an existing design problem by gradually rewording it into a broader, more abstract statement. There is not one answer; you can repeat this process multiple times to come up with variations. Use the provided template (p.182) to guide you.

1 **Write a one-sentence description** of the product or service you are designing. Try to include the following three things:
- A user: one person or a group of people who will use it.
- A setting: where it takes place.
- A goal: what it should do.

E.g. if we were designing a circus show we may describe a circus as follows: a circus is a cheap travelling show to entertain children (Seelig, 2012).
[5 minutes]

2 This first description often contains a lot of assumptions. Start to shift your frame by **changing keywords** relating to the person, setting and goal. Where possible make them less specific and more abstract.
E.g. change user: a circus is a cheap travelling show to entertain **everyone**.
E.g. change setting: a circus is an **experience** that entertains **everyone**.
E.g. change goal: a circus is an **experience** that **captivates and immerses everyone**.
[10 minutes]

3 To improve your statement, **look for the underlying goals**. Ask yourself "why" people use your product, service or system, and try to summarise the reason succinctly.
E.g. people go to a circus to step outside of ordinary life and bond with friends and family through a shared experience and time spent together.
[10 minutes]

4 **Try to rephrase your problem statement** to include these underlying goals. Remove the original noun if you started with one.
E.g. design an experience that captivates and immerses everyone who sees it and helps them step outside their ordinary life through a shared moment.
[5 minutes]

5 **Use your new problem statement** to generate ideas by methods, such as brainwriting 6-3-5 (p.28) or bodystorming (p.26).

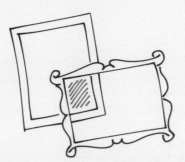
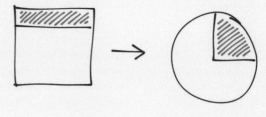

Research Visualisation

Giving your research pride of place on the wall

ACADEMIC RESOURCES:

Abram, S., Popin, S., & Mediati, B. (2016, May). Current States: Mapping Relational Geographies in Service Design. In Service Design Geographies. Proceedings of the ServDes. 2016 Conference (No. 125, pp. 586-594). Linköping University Electronic Press.

Stappers, P. J., van der Lugt, R., Sleeswijk Visser, F., & van der Lelie, C. (n.d.). RichViz! Inspiring Design Teams with Rich Visualisations.

The purpose of design research is to gain valuable insights through data. By analysing and synthesising this data, we are able to derive information that confirms, informs and inspires. These insights help us move forward with a design project, with confidence that we understand our users' needs. It's important to capture research insights in a way that is easy for everyone to understand. Often this is done through generating a research report. However, it can be difficult to quickly gain an overview of the design research findings from a lengthy report.

Research visualisations capture the most relevant results of a research process and display them in a prominent and visual manner. Often they take the form of a large-format poster that can be hung on the wall and used as a point of reference. They create sustained engagement within the design team and ensure that the research lives on throughout the design process. The objective of research visualisations is to transfer the inspirational qualities of the original research, demonstrate an understanding of the main issues and show empathy with the target audience.

Research visualisations often use the principles of infographics and other visual communication techniques and apply different frameworks to present information. Common frameworks include concept maps, characters, and chronological or spatial models. As with any infographic, the most suitable framework and format depends on the kind of information being communicated. There are no hard rules, but many best practice guidelines. Making a good research visualisation is a creative process in itself. Documented techniques for creating research visualisations include 'Rich Visualisation' (Stappers et al., n.d.) and 'Current State Maps' (Abram et al., 2016).

EXERCISE

YOU WILL NEED
Pencils, coloured pens, A3 paper,

In this exercise, you will create a communicative research visualisation using a set of data as inspiration. Focus on your own design problem, or follow the 'Museum Visitor Experience' brief (p.142).

1. **Make a selection of which aspects of your research to communicate.** Start by identifying key themes. Choose important quotes or insights that support these themes and help the reader to empathise with the user and their situation.
[5 minutes]

2. **Choose a visual framework.** Look for a visual metaphor that helps you to express your findings – this will form the basis of your research visualisation. You can also use a combination of metaphors.
E.g. a timeline from morning to night, or Monday to Friday
E.g. a drawing of a physical space, such as a map or floor plan
E.g. expression using one or more characters
Relations can be shown by drawing lines between themes, or providing other suitable divisions on the page – the possibilities are endless. The right visual framework to choose will depend on your research outcomes.
[10 minutes]

3. **Start by sketching the most important elements.** What is the first thing readers will see when they look at your poster from a distance? The most dominant visual element should also confer the most important information. Incorporate a catchy title and a summary.
[10 minutes]

4. **Focus on adding the different themes.** Think about what the visitor should see when they examine your poster for a bit longer. Important themes should stand out. They should be heavily supported by relevant visuals; at least one for each important message you want to communicate.
[10 minutes]

5. **Add details to support the information.** This is where you add detailed arguments, supporting quotations and other evidence. The user's voice should be represented where possible to help people empathise with the target audience. Examining your research visualisation extensively should give people access to these details.
[20 minutes]

6. **Finalise the documentation** by creating a second iteration on paper, or using software for a more professional result.

Role-playing

Exploring the perspectives of users

ACADEMIC RESOURCES:

Brandt, E., & Grunnet, C. (2000, November). Evoking the future: Drama and props in user centered design. In Proceedings of Participatory Design Conference (PDC 2000) (pp. 11-20).

Svanaes, D., & Seland, G. (2004, April). Putting the users center stage: role playing and low-fi prototyping enable end users to design mobile systems. In Proceedings of the SIGCHI conference on Human factors in computing systems (pp. 479-486). ACM.

The role-playing method is used to evaluate an existing design or a design concept represented as a prototype. It is based on role-playing in theatre, where actors take on the role of a particular character and subsequently act out scenes. In a design context, the actors can be members of the design team, the extended project team, prospective users, other stakeholders, or even trained theatre actors. They are asked to step into the shoes of a character and to act out scenes from that character's point of view, which involve interactions with the existing design or prototype.

The roles are prepared in advance before being assigned to actors and are recorded on role cards, which include the character's goals, motivations, tasks and so on. Props (see bodystorming, p.26) or empathic modelling suits (p.56) can be used to mimic specific aspects of the character's experience, such as having arthritis or poor eyesight.

Experience prototyping (p.58) works well for representing products or other objects used in the role-play.

Roles can be visually depicted by writing them on a Post-it note and attaching it to the actor using simple terms, such as 'cashier' and 'elderly shopper'. Actors can also take on the role of objects, such as an electric car or a smartphone, and act out dialogues between a user and those objects. These roles should be carried out by members of the design team.

Role-playing might seem silly at first, but it is a highly effective tool for getting quick feedback on early design concepts, as it reveals issues that may have been overlooked or require further design work to be resolved.

EXERCISE

YOU WILL NEED
3–4 people, paper, pen, props (optional)

In this exercise, you will use role-playing to evaluate a prototype or existing product by assigning your group members different roles and acting out scenarios of use. Use the provided template (p.183) to write your own role cards, or use the resources on the companion website.

1. **Give each person a role card** or write your own roles specific to the prototype or product you are exploring. If you don't have your own prototype, choose an existing product or situation:
   E.g. using an ATM to withdraw money
   E.g. waiting in the hospital waiting room
   E.g. visiting a theme park
   Add details to your prototype/product scenario, using Post-it notes or other visual cues to clarify the time and place.
   [5 minutes]

2. Get your group members to **familiarise themselves with their role descriptions** by reflecting upon their character's needs and motivations. This is a chance for them to ask for any clarifications.
   [5 minutes]

3. **Instruct each group member to interact with the product or prototype** from the perspective of their given role. They should:
   - Walk through all the steps that might be taken in a typical scenario of use. These may be sequential, simultaneous or overlapping.
   - Try to explore all the most likely occurrences between that character and the product.
   - Improvise when encountering other characters or parts of the system.
   [15 minutes]

4. Ask your group members to **express their thoughts** and feelings using think-aloud (p.124) while interacting. Record any interesting observations about their actions or words.

5. When all group members are finished, **ask them to provide feedback** from the perspective of their role. How well did the experience match their needs? How did it satisfy or frustrate them? Note the strengths and weaknesses of the design.
   [5 minutes]

6. **Use this input** to list suggested changes to the design to make it more accommodating to the chosen user roles.
   [2 minutes]

Design. Think. Make. Break. Repeat.

Scenarios

Exploring designs through storytelling

ACADEMIC RESOURCES:

Brooks, K., & Quesenbery, W. (2011). Storytelling for User Experience. Rosenfeld Media.

Design Scenarios - Communicating the Small Steps in the User Experience, Interaction Design Foundation Website. [Link: https://www.interaction-design.org/literature/article/design-scenarios-communicating-the-small-steps-in-the-user-experience] Design Scenarios - Communicating the Small Steps in the User Experience (2017, July 1). The Interaction Design Foundation. Retrieved from https://www.interaction-design.org/literature/article/design-scenarios-communicating-the-small-steps-in-the-user-experience

Rosson, M. B., & Carroll, J. M. (2009). Scenario based design. In Human-computer interaction: Development process (pp. 1032-1050).

Scenarios – also referred to as future scenarios or user stories – use storytelling to explore design ideas by grounding them in a real context. They connect design ideas to the people who will use the product or service. Their focus, therefore, is on the relationship between people and the designed product or service, their objectives, the context and potential social implications.

Scenarios can be used for documenting and communicating ideas and design concepts without representing them physically. Sometimes, scenarios are also used for describing how people currently use an existing product or service, in order to highlight issues with current solutions.

A typical scenario is a short fictional story written about lead characters performing certain tasks and interacting with a proposed design concept. The characters represent the user(s) as well as other people who are associated with using the designed product. For example, a scenario may involve the steps taken by the primary user of a mobile application to capture a picture. This could be expanded to include how that picture is shared with their social network and the ways in which their social interactions are influenced as a result. A simplified form of scenarios is to record interactions using the following structure: 'As a (role) I want (something) so that (benefit).'

Scenarios are often developed following an ideation method, such as brainstorming, and rely on predetermined information collected about the intended users of the design concept, for example, in the form of personas (p.100).

EXERCISE

YOU WILL NEED
Pen, paper

In this exercise, you will write a scenario where a specific user interacts with a specific product or service, ideally in the form of a new concept that you have generated. Use the provided template (p.184) to guide you.

1 If you don't have a **design concept and user representation**, e.g. a persona (p.100) or extreme character (p.62), use the resources on the companion website.
[5 minutes]

2 **Discuss the user** in your scenario. Who are they? What do they care about? Why might they need to use the product you have in mind?
[5 minutes]

3 If creating a scenario for an existing product, **make a list of reasons why the product may not work** for this user. If you are creating a scenario for a new design, **make a list of features of the design** that you want to showcase. These lists contain details that you should include in your scenario.
[5 minutes]

4 Use the work template to **draft the main narrative of your scenario**, covering the following details in your story:
- Set the scene. What is the context of use?
- *E.g. 'It was Friday 5pm ..."'*
- Introduce the character. What motivates them to use the product?
- What actions will they take to reach their goal? Think of the important steps.
- Do they interact with other people along the way? Add relevant characters.
- How does the story end?

[10 minutes]

5 Work on your story, choose a title and **write the narrative**.
[10 minutes]

6 **Identify three key qualities** for your design concept that emerged from the story. You may go a step further by creating a storyboard (p.120) to visualise the scenario.
[5 minutes]

Science Fiction Prototyping

Using the future to improve the now

ACADEMIC RESOURCES:

Dourish, P., & Bell, G. (2014). Resistance is futile: reading science fiction alongside ubiquitous computing. Personal and Ubiquitous Computing, 18(4), 769-778.

Johnson, B. D. (2011). Science fiction prototyping: Designing the future with science fiction. Synthesis Lectures on Computer Science, 3(1), 1-190.

Shedroff, N., & Noessel, C. (2012). Make It So: Interaction Design Lessons from Science Fiction. Brooklyn, New York, USA: Rosenfeld Media.

Science fiction prototypes are stories placed in the distant future. They allow the fictional exploration of scenarios, in which people interact with envisioned products or services. The narrative of the story is based on real scientific principles and technologies, but explores their use in an environment that is free of constraints. The story follows a set structure, which includes identifying the characters, the scientific principle or technology, and so on. Critically, the narrative should include an inflection point, possibly leading to a disaster, as well as an exploration of the implications and how the characters can recover from or overcome this disaster.

Once a story narrative is developed, it is turned into a prototype representation of how an envisioned product or service would be used in the future. This is often an essay, comic or movie. However, even the skeleton of the narrative can be a useful artefact in a design process. The science fiction prototype can then be used to reflect on which of its elements could be brought back into the current design situation. The method is used for speculative prototyping as well as ideation – by using elements from the science fiction prototype to inform the design of a solution.

Science fiction prototyping is used by tech companies as a way to explore how their technology will be used in future scenarios. For example, Intel uses this method to determine how people will be interacting with semiconductor-based products in the future, which helps them to identify requirements for the development of new semiconductor technology. Science fiction prototypes are also useful to communicate speculative ideas and scenarios within design teams.

EXERCISE

YOU WILL NEED
Pen, paper

In this exercise, you will develop a narrative science fiction prototype using the provided template (p.185). Focus on your design problem, or choose the 'Autonomous Vehicles' brief (p.140).

1 Based on your chosen brief, **pick a scientific principle or technology and build a fictional world** around it. Include an explanation of what it is and how it fits into the world you are creating. Develop the characters in your story and the locations where the action will take place. Record your ideas in the template using notes or bullet points.
[10 minutes]

2 **Introduce the science or technology** into the narrative of your story. This step is called the scientific inflection point. Again, use notes or bullet point form to explore this.
[5 minutes]

3 **Explore the implications and ramifications** your science or technology has on the world you created. Does it affect people's lives for better or worse? Is there a risk that it might lead to a disaster or even the end of the world as we know it? This step is referred to as the ramifications of the science or technology on people.
[10 minutes]

4 With the science or technology now part of the future scenario, **describe what happens next**. If there was a disaster, how could it be fixed to save the world? Does the science or technology need to be modified? This step is referred to as the human inflection point.
[10 minutes]

5 **Develop your outline** into a full science fiction story, if you have time to do so. Otherwise use the outline for ideation purposes in step six.

6 **Reflect on what you learned** from creating the outline of your science fiction story. What are the possible implications, solutions or lessons? What are aspects that could be taken into the current reality and integrated into an envisioned solution that addresses your chosen brief?
[10 minutes]

Design. Think. Make. Break. Repeat.

Service Blueprints

Documenting the visible and the invisible

ACADEMIC RESOURCES:

Shostack, G. L (1982) How to Design a Service. European Journal of Marketing, Vol. 16 Issue: 1, pp.49-63

Stickdorn, M., Schneider, J., Andrews, K., & Lawrence, A. (2011). This is service design thinking: Basics, tools, cases. Hoboken, NJ: Wiley.

Whenever we take part in a service, we only see the part that is happening 'on stage'. Entering a cafe, we might choose a muffin, pay the cashier and enjoy our food while reading the paper. But even a service as simple as this consists of both visible and invisible components. Much needs to happen 'backstage' – muffins get baked and delivered, the electronic payment system transmits payment information, and the cafe gets stocked with newspapers each day.

Service blueprinting plots out all these different elements in order to form a picture of the overall system. This helps to make it more comprehensible, and identify strengths and weaknesses. The service blueprint resembles a flowchart in which the horizontal dimension shows progress through time, and connections between steps are represented by arrows. A line in the middle of the page called the 'line of visibility' helps to indicate what is apparent to the user and what is hidden.

Everything above the line is 'on stage' and everything below the line is 'backstage'.

Revealing this complexity allows us to understand an experience in detail, which can help us to make improvements or design new services. The technique can be used for developing ideas in the early phases or for testing current and proposed solutions. By charting the user's role within a wider system, designers can describe and understand how each individual component is connected. In advanced versions of service blueprinting different shape codes are used for each element of the system to help indicate different kinds of activities.

EXERCISE

YOU WILL NEED
Pen, paper, a partner

In this exercise, you will plot all the steps of a system onto a service blueprint, including potential failure points. Use the template on the companion website to support you.

1. **Choose a service to plot** out, or focus on the experience of taking a train.
[3 minutes]

2. **Write down five key steps of the service** as experienced by the user. Write these in the five numbered boxes across the template.
E.g. the first stage of train travel might be 'planning the journey'.
[10 minutes]

3. For every step, **try to identify at least one corresponding step** that takes place behind the scenes. These actions are required, but the user won't see them. Write these in the boxes below the 'line of visibility'.
E.g. for the user to plan their journey, they need to retrieve up-to-date timetable information from a database.
[10 minutes]

4. **Identify additional corresponding steps.** Sometimes multiple things may happen below the line of visibility to facilitate a single user action. Add as many boxes as you need to represent these.
E.g. the correct train platform is indicated by a board and by a PA announcement.
[5 minutes]

5. **Connect the steps in chronological order**, to show the flow of information through the system. Arrows can pass across the line of visibility and back again.
[5 minutes]

6. **Add failure points**, indicated by an F in a circle, and use these to lead into an alternative version of the experience. Try to think of different logical ways the system might fail.
[5 minutes]

7. **Redesign the service using a new blueprint**. Discuss your blueprint with a partner. Where can the system fail easily? What parts seem convoluted with many arrows? Is there a lot going on behind the scenes while the user is left waiting? Take these factors into account as you design a new improved service.
[30 minutes]

Sketching

Communicating and thinking through pen and paper

ACADEMIC RESOURCES:

Buxton, B. (2010). Sketching user experiences: Getting the design right and the right design. Morgan Kaufmann.

Schön, D. A. (1984). The architectural studio as an exemplar of education for reflection-in-action. Journal of Architectural Education, 38(1), 2-9.

Sketches are tools used by designers to visually represent and depict physical aspects while composing an idea for a product or service. The act of sketching has a series of advantages. Sketches are cheap and quick to materialise, therefore, easily disposable and replaceable. Because of this, it is possible to articulate several ideas without having to compromise valuable resources, such as time or expensive materials. Rather than trying to verbalise what ideas are going to look like, they can be quickly represented through a sketch. By putting ideas on paper, they become clearer, facilitating a generative dialogue between your thoughts and your sketch. The act of sketching creates a 'back talk' (Schön, 1991), allowing for new meanings to emerge that can be explored through further sketching. As a result, ideas might end up being more comprehensive and creative.

Sketching only requires pen and paper. It can be used in a collaborative setting to collectively explore and sketch out ideas. It also forms the foundation for other methods such as sketchnoting (p.118) and storyboarding (p.120). Although the flexibility of sketches facilitates their use throughout different stages of the design process, they are generally more relevant during the idea generation phase.

Sketches are also excellent tools to communicate ideas to stakeholders and colleagues. The idea of sketching and drawing can be quite overwhelming for people with no training in design or art, yet it is possible to elaborate good sketches by using simple resources, such as basic shapes and lines. The effectiveness of sketches is neither linked to their artistic merit nor the need to conform to traditional ideas of 'beauty'. Successful sketches facilitate discussion, understanding and critique.

EXERCISE

YOU WILL NEED
Pen, paper

In this exercise, you will become familiar with sketching and develop your own sketch vocabulary (Buxton, 2010). These exercises don't require any drawing skills and can be used by novice and expert designers alike. Use the provided template (p.186) to get you started.

1. Warm-up with the door sketching exercise using the provided template. This will help you to get a feel for sketching quickly and intuitively without getting hung up on the details.
[5 minutes]

2. Sketch objects in the room around you. Try to reduce each one to a minimal and simple expression of that object – your sketches don't need to be detailed illustrations.
E.g. pens, paper, books, laptop, stapler, cup, table, chair, mouse, USB stick, notebook
[10 minutes]

3. Sketch emotions. Draw different people and their facial expressions, representing a variety of different emotional states.
E.g. surprised, annoyed, confused, busy, chatting happily with someone else, having a 'eureka' moment, relaxed
[10 minutes]

4. Sketch people carrying out different tasks. Use the emotional states that you just practised to create characters that are expressive.
E.g. working in the office, reading a newspaper, biting an apple, riding a bicycle, giving a public presentation, walking a dog, eating sushi
[10 minutes]

Sketchnoting

Documenting processes through sketching

ACADEMIC RESOURCES:

Rohde, M. (2015). The Sketchbook Workbook. Peachpit Press.

Sketchnotes are visual notes that are a popular technique for capturing the content of a public talk or panel discussion through an illustration combining sketches and annotations. In design, sketchnotes are used to capture a current process, make a plan, visualise a design idea, or communicate within the design team and to stakeholders. Similar to sketching (p.116), sketchnoting can also be used for ideation. Creating a visual note of an idea for a concept allows for reflection on the specifics, values and implications of the concept. Compared to other methods, this has the advantage of being able to use visual cues to access tacit aspects of an idea without sacrificing detailed content.

Sketchnoting merges sketching and annotations to further emphasise the meaning of what is discussed and to give a visual hierarchy to existing ideas. Unlike mind maps (p.88), sketchnotes are self-explanatory and easier to follow as they are intended to be used as a way to communicate a message. It is important to keep this communication aspect in mind when creating sketchnotes, and, if possible, to evaluate the sketchnote by discussing it with other team members.

Sketchnoting uses specific visual vocabulary and elements in addition to texts and actual drawings, such as symbols, arrows, boxes and different typographies. These graphical elements are useful starting points, but they are not intended to limit the visual style of the note. The barrier to sketchnoting is lower compared to other sketching techniques as it offers familiarity with common ways of organising content, such as taking notes or creating lists.

EXERCISE

YOU WILL NEED
A partner, pen, paper, coloured pencils, highlighter

In this exercise, you will practice sketchnoting by describing a process in the format of a sketchnote. For example, describe the steps of cooking your favourite recipe, explain how you prepare to travel abroad or record your morning routine.

1. **Make a list** of the logical steps that are needed to perform the chosen activity. Note down each step, trying to be as detailed as possible in your description.
[5 minutes]

2. **Reflect** upon the information, and highlight the parts of the text that seem more relevant. Think of your audience and how to get your message across concisely.
[2 minutes]

3. **Consider the most logical visual organisation**. Left to right? Top to bottom? Or maybe from the centre to the sides? Think about other more expressive ways of organising the content. Keep in mind that clarity is key.

4. Take some time to **write the title** in detail. Pay attention to the typography you are using and aspects such as size and location of your text.
[2 minutes]

5. **Start sketchnoting using sketches, symbols and annotations** to describe the process. Take advantage of the expressiveness of visual tools, to illustrate the aspects of your process that cannot easily be described through words.
E.g. you need a specific kind of pan for your recipe. Using words to explain what the pan looks like might be difficult, but a clear sketch can easily illustrate which kind of tool is required.
[10 minutes]

6. **Get feedback from your partner** to evaluate whether your sketchnote effectively communicates the process. Is your partner able to understand the described process? Which parts are not quite clear yet? What aspects work well? Ask questions.
[5 minutes]

7. **Review your partner's sketchnote**, highlighting strengths and weaknesses of each visual solution.
[5 minutes]

Design. Think. Make. Break. Repeat.

Storyboarding

Using the power of comics to explain concepts

ACADEMIC RESOURCES:

Davidoff, S., Lee, M. K., Dey, A. K., & Zimmerman, J. (2007, September). Rapidly exploring application design through speed dating. In International Conference on Ubiquitous Computing (pp. 429-446). Springer Berlin Heidelberg.

Greenberg, S., Carpendale, S., Marquardt, N., & Buxton, B. (2011). Sketching user experiences: The workbook. Elsevier.

Truong, K. N., Hayes, G. R., & Abowd, G. D. (2006). Storyboarding: an empirical determination of best practices and effective guidelines. In Proceedings of the Designing Interactive systems (pp. 12-21). ACM.

Storyboards in design are used to visually explore the interactions between people and products or services. They can either represent an existing situation or communicate an envisioned situation. When depicting existing situations, the story should be based on real data, for example, collected through contextual observation (p.44). Storyboards of existing situations are effective for highlighting issues with current experiences. Storyboards of envisioned situations can be used for evaluating early concepts with other team members or prospective users and for communicating concepts to others.

Storyboards can be either hand-drawn or digitally composed illustrations that take techniques from film-making and comics. They consist of rectangular frames arranged horizontally or vertically in temporal order to narrate a story. Each frame represents a 'shot', similar to the use of storyboards in film. Speech and thought bubbles are used to represent dialogues and thought processes. To keep the story easily accessible, the number of panels should be between three and six. If more panels are needed, they can be included as an additional storyboard. Details in a panel are used to focus the viewer's attention on the important parts of the scenario, such as one of the characters interacting with a product. Descriptions above or below each panel are used to explain the scene within the panel. Time can be indicated either explicitly using a clock or calendar, or through implicit indicators such as a rising sun or contextual dialogues.

The characters in the story should be based on user representations, for example, in the form of personas (p.100) or extreme characters (p.62). Characters can interact with each other as well as the explored product or service to express emotions and relationships.

*FADES TO WHITE

EXERCISE

YOU WILL NEED
Paper, pens, coloured pencils

In this exercise, you will create a storyboard documenting an existing situation or demonstrating a new design idea. Use the provided template (p.187) to get you started.

1. Reflect on the user of the product or service you have in mind. If you need a topic, you can use a sample persona from the companion website and focus on one of the following:
E.g. getting money out of an ATM
E.g. purchasing concert tickets
E.g. making a cup of coffee
[3 minutes]

2. Write down three to five key steps that the user would go through when interacting with the product or service. Plan what 'shots' and techniques you could use to illustrate these steps. Shots can include:
- Wide shot: showing the surrounding context
- Long shot: showing a person with their body fully visible and his/her surrounds
- Medium shot: showing a person's head and shoulders
- Over-the-shoulder shot: looking at things 'over the shoulder' of a person
- Point of view shot: showing things through a person's eyes
- Close-up shot: showing a detailed view of a device or interface.

[5 minutes]

3. Draw your storyboard in the template. Try to **begin with a 'wide shot'** to establish an impression of where the story begins and to introduce the objects or people that are important.
[5 minutes]

4. For each remaining step illustrate what the person would do. You can just use simple symbols and stick figures. Use a variety of shots to show relevant parts of the environment and the interactions between the person and the evaluated product or service.
[15 minutes]

5. Add short captions to describe each step. Ideally, every panel should show a single action accompanied by a sentence explaining the action. To improve your storyboard try the following:
- Use bold outlines or highlight colours to draw attention to important parts.
- Use arrows to indicate important directions of movement.

[5 minutes]

Design. Think. Make. Break. Repeat.

Thematic Analysis

Finding patterns that make sense of what people say or write

ACADEMIC RESOURCES:

Braun, V., & Clarke, V. (2006). Using thematic analysis in psychology. Qualitative research in psychology, 3(2), 77-101.

Boyatzis, R. E. (1998). Transforming qualitative information: Thematic analysis and code development. Sage.

Saldaña, J. (2009). The coding manual for qualitative researchers. Thousand Oaks, California: Sage.

Whether using interviews (p.78), focus groups (p.64) or written responses in questionnaires (p.102), what people say or write is important. However, in its raw form, this data is difficult to make sense of so that it can be fed back into the design process. Thematic analysis is a structured method for analysing, interpreting and managing qualitative data in a way that allows us to navigate from data to ideas.

Thematic analysis focuses on establishing themes. Themes are defined as short words or phrases that assign 'a summative, salient, essence-capturing and/or evocative attribute' (Saldaña, 2009, p.3) to a portion of written or visual data. For example, interviewees may repeatedly express their opinion on the price, cost or value of a certain product; though each interviewee may use different words, phrases or tones of voice, all of these expressions could be grouped under a single theme 'cost sensitivity'. The theme may relate to other data, for example, we might choose to catalogue high or low-cost sensitivity, relate it to other themes, such as 'quality', 'lifestyle' or 'income'. Thematic analysis can be used for bottom-up as well as top-down interpretation of data. Following a bottom-up approach, the themes emerge from engaging with the data. When using a top-down approach, the themes are predetermined, for example, based on a research question or hypothesis that acts as starting point for the analysis.

When identifying a theme it is important to remember that qualitative data is often messy, complicated and open to interpretation. Despite these challenges, exploring data in this way helps us to connect with people and empathise with their experiences, so that we can ensure that these remain at the centre of the design process.

EXERCISE

YOU WILL NEED
A partner, pen, highlighter, paper

In this exercise, you will perform thematic analysis of a set of interview data using a bottom-up approach. The data could be from your own interviews, or you can use resources on the companion website.

1 Each person should **read a section of the interview data**. Highlight and annotate the text as you go. Start by reading with an open mind. Focus on what the data is saying and avoid trying to find what you want or expect. Look for the following:
- What topics occur frequently?
- What topics are accompanied by emotional responses?
- Which words are used?

[10 minutes]

2 **Create your first set of themes**. Each theme should be a word or statement that groups two or more topics the interviewees discussed. Record these themes in the thematic analysis template.

[10 minutes]

3 **Group the data according to your themes**. Record the number of people who express that theme, the total number of references, and reflective notes:
E.g. 73 references to cost sensitivity throughout all interviews
E.g. cost sensitivity was a theme in 10 out of 15 interviews

[20 minutes]

4 **Open up the analysis by using alternative lenses to explore the data**. Each person should select one of the following provocation points and look for additional themes that may emerge:
- Issues, needs, motivations, etc.
- Identity: gender, socioeconomic status, age, etc.
- Emotions and attitudes: likes, dislikes, perceptions, etc.
- Behaviour: usage, habits, tools, etc.

Re-read your interview data and add additional themes to the template.

[10 minutes]

5 **Discuss, combine and group your findings in pairs**. What did you think was important and why? How do the different themes relate to one another? There is no single right answer, just different ways of seeing the data.

[5 minutes]

6 **Consolidate your analysis in a diagram or drawing** that shows how themes relate to each other. Share this with your design team.

[5 minutes]

Think-aloud Protocol

Learning from listening to your users' thoughts

ACADEMIC RESOURCES:

Ericsson, K. A., & Simon, H. A. (1993). Protocol analysis. Cambridge, MA: MIT press.

Nisbett, R. E., & Wilson, T. D. (1977). Telling more than we can know: Verbal reports on mental processes. Psychological review, 84(3), 231.

Rooden, M. J. (1998). Thinking about thinking aloud. Contemporary Ergonomics, 328-332.
Comprehensive, but not easily accessible

Various methods and tools exist for testing the design of an existing or new product or service, such as usability testing (p.126) and contextual observation (p.44). It can be difficult, however, to gain a deep understanding of users' interactions, and the reasoning that led to those interactions, purely from observing people interacting with a product or service. Methods such as interviews (p.78) and questionnaires (p.102) address this to some extent by asking people to verbalise their experiences. However, doing this retrospectively makes the accuracy and completeness of the information provided reliant on the user's memory and ability to recall their interactions.

The think-aloud protocol method allows designers to gain access to people's thought processes while they are interacting with a product or service. It is often used in combination with usability testing to help us find out whether people understand and can use a product or service. It encourages people to verbalise what they are thinking as they perform a task, revealing their cognitive processes. This can help reveal the gap between the designer's and the user's mental models, and provide insights into how users are actually experiencing the product or service in a real or simulated scenario.

Talking out loud while using a product or service can feel very unnatural for some participants, which might potentially affect the validity of the results. A variation that addresses this is 'co-discovery', in which two people interact with a product or service at the same time, leading to more natural conversations about their experience.

EXERCISE

YOU WILL NEED
A partner, paper, pen, audio recorder (optional)

In this exercise, you will use the think-aloud protocol method for evaluating an existing prototype to gain insights about what is actually happening while users interact with the product. Use the provided template (p.190) to record your data.

1. **Begin with a prototype** that you want to test. If you don't have one, choose a product or service that you have readily available.
   E.g. your smartphone, your email inbox or your dishwasher

2. Give your participant **a task to perform**, which will take them through the critical steps of using the product or service.
   E.g. locate the GPS function and look up the nearest train station.
   E.g. find all the emails that were received on the 17th of October last year.
   E.g. pack the dishwasher and set the cycle to wash a heavy load.
   [10 minutes]

3. **Instruct your participant to describe the steps** they are taking and express their thoughts and feelings as they engage with the task.

4. For each step of the task, **listen and note down** what the participant is saying – this is known as the verbal protocol. To help you record things quickly, use the think-aloud data collection form (p.190). With your participant's consent, you can also use an audio recorder so that you can review the data later on.

5. If your participant stops talking while focusing on the task, gently **encourage them to keep talking**. Do not explain any part of the interface to the participant – otherwise, you will get less value out of the test.

6. **Note down the strengths and weaknesses** of the design from the perspective of your test user. How well does it help them achieve their goal? What issues frustrate them or prevent them from performing the task effectively?
   [2 minutes]

Design. Think. Make. Break. Repeat.

Usability Testing

Identifying design flaws by testing early and often

ACADEMIC RESOURCES:

Nielsen, J. (1994). Usability engineering. Elsevier.

Rubin, J., & Chisnell, D. (2008). Handbook of usability testing: how to plan, design and conduct effective tests. John Wiley & Sons.

Sauer, J., Seibel, K., & Rüttinger, B. (2010). The influence of user expertise and prototype fidelity in usability tests. Applied ergonomics, 41(1), 130-140.

The person that designs a product will always have a much deeper understanding of how it is supposed to work. For a design to be successful, the designer's mental model needs to match the user's mental model. People form their mental model based on the design and interface of the actual product as well as their previous experiences and cultural background.

Usability testing allows us to determine whether people understand how to use a product, in other words, whether the conceptual model of the product matches the user's mental model. More specifically, usability refers to efficiency, effectiveness and satisfaction (Nielsen, 1994). Usability testing should measure all three aspects. For example, efficiency can be measured by recording the time it takes for a user to complete a certain task. Effectiveness can be measured by recording whether people were able to complete a task. Satisfaction can be measured through recording quotes from your participant, either while they are interacting with the product using the think-aloud protocol (p.124) or in a post-test interview or questionnaire (p.102).

Usability testing is typically done in a controlled environment either in a dedicated lab or an office space. To achieve valid results, it is important to ensure that participants feel comfortable. It is recommended to test with at least five participants, which usually reveals 80 percent of the issues in a design. Usability testing can be completed with a prototype or an existing product. A good approach is to do several rounds of testing with various iterations of a prototype.

EXERCISE

YOU WILL NEED
A partner, audio recorder (optional), stopwatch, pen, paper

In this exercise, you will test the design of a product with a prospective user representing your target audience. You can use an existing product (e.g. a social networking app) or a low-fidelity prototype (p.84) produced in a previous exercise. Use the provided template (p.191) to take notes.

1. **Identify three or more tasks** related to features you want to test. Add any details needed to complete the task.
   E.g. task: searching for a person. Task formulation: 'You would like to add a person. His name is Peter Friend.'
   Don't provide specific instructions, such as which button to click on. You want to find out whether people can complete tasks without help.
   [10 minutes]

2. **Prepare pre-test and post-test interview questions**. Pre-test questions should include factors relevant to your product.
   E.g. previous experience with similar products, demographic data (age, profession)
   Post-test questions should capture the participant's experience with the product.
   E.g. what they liked, what can be improved
   You can use the System Usability Scale (SUS) questionnaire template (p.192).
   [10 minutes]

3. **Explain the product and purpose** of the test and that you are testing the product, not the participant. Outline the procedure and ask for your participant's permission to audio-record the session. Ask your pre-test questions and record the answers.
   [5 minutes]

4. **Give your participant written instructions for the first task**. Using the work template, record start and end time for each task, any interesting observations and the number of mistakes they make. If the participant can't complete a task alone, you can provide help but be sure to record this. Using the think-aloud protocol method (p.124) can give additional insight. Continue with the remaining tasks.
   [30 minutes]

5. **Ask your post-test questions** and record the answers. Finish the test by thanking your participant.
   [10 minutes]

6. **Prepare a report** about your findings. Include details about the procedure (number of participants, tasks, set-up, etc.). This is an important step in design projects, as the report is used to inform the next iteration of the product.
   [several hours]

Design. Think. Make. Break. Repeat.

User Journey Mapping

Understanding complex user experiences

ACADEMIC RESOURCES:

Nenonen, S., Rasila, H., Junnonen, J. M., & Kärnä, S. (2008). Customer Journey – a method to investigate user experience. In Proceedings of the Euro FM Conference Manchester (pp. 54-63).

Stickdorn, M., Schneider, J., Andrews, K., & Lawrence, A. (2011). This is service design thinking: Basics, tools, cases. Hoboken, NJ: Wiley.

A user journey map is a way of presenting and describing every important step of a complex experience in a single overview. It includes what the user is doing, thinking and feeling, and also indicates what physical or technological infrastructure is supporting them along the way. User journey maps are useful when designing complex experiences like banking, healthcare or public transport.

For example, a person who is opening a new bank account can do this in many different ways. They could go into the bank in person, visit the online website or phone the bank. The bank has to be ready to offer a seamless experience no matter which of these avenues the user chooses. Information must be carried over from every step to the next in a logical way and stay consistent, even if the user switches from one type of interaction to another. The user needs to understand where they are in the process and what they have to do next. Keeping an overview of this complexity while designing the details of individual interaction points can be challenging.

The user journey map facilitates this process by visually representing all steps of an experience in a table-like format. Columns are used to describe the stages or phases of an activity. Rows are used to describe the various dimensions of a journey, such as thoughts, emotions, goals, touchpoints, pain points and opportunities. The dimensions should be chosen to align with the project and its objectives. By mapping user's experiences across the dimensions for each stage of the journey, it is possible to gain an understanding of the whole experience. This gives clarity about which "touchpoints" need to be designed or redesigned in order for users to achieve their goals with as little friction as possible.

EXERCISE

YOU WILL NEED
A pen

In this exercise, you will create a user journey map using the template (p.193). Before you start, you will need a good understanding of the current user experience of a product or service, based on prior research. Follow the 'Supermarket of the Future' brief (p.143) for this exercise.

1 **Choose whose journey it is.** Select a user and the experience to be examined, or choose a persona from the resources on the companion website, and map their journey as they try to achieve a specific goal.
E.g. doing the grocery shopping for the week
[10 minutes]

2 **Plot out the main stages of the experience**, in the top row of the template. The experience may begin earlier than you think; usually there is some planning activity on the part of the user. Write one main phase in each column.
E.g. for the grocery shopping example, visitors will have a planning phase, a shopping phase, and a purchasing phase – as a minimum.
[15 minutes]

3 **Write down the activities** carried out during each of the main stages you identified, in the second row of the template.
E.g. writing down a shopping list, checking current inventory in the fridge and pantry, and gathering re-usable shopping bags may all be part of the planning phase.
[15 minutes]

4 **Devote the next row to the user's emotions**, and note down how the user is **thinking and feeling** during the different stages and activities.
[10 minutes]

5 **Use the next row to record the touchpoints** required to complete these activities - where do each of these things happen?
E.g. checkout station, website, supermarket personnel, deli counter
[15 minutes]

6 **Write down all the pains (what is negative) and the gains (what is positive)** about the experience across the selected stages. This provides a focus for potential future redesigns – your user journey map can be used as a tool to analyse where the current experience can be improved.
[10 minutes]

Design. Think. Make. Break. Repeat.

User Profiles

Describing your users' key attributes

ACADEMIC RESOURCES:

Courage, C., & Baxter, K. (2005). Understanding your users: A practical guide to user requirements methods, tools and techniques. San Francisco: Morgan Kaufmann Publishers.

Holtzblatt, K., Wendell, J.B., & Wood, S. (2005). Rapid Contextual Design: A How-to Guide to Key Techniques for User-Centered Design. Elsevier Inc.

When designing a product or service we need to design for different users. Though some products and services may target a more specific user group (e.g. surgeons) than more general user groups (e.g. mobile phone users), all user populations contain a degree of variation. User profiles are a way of mapping this variation.

User profiles present key attributes of different identified users in an overview table that can show the full spectrum of possible users; on one axis are the names of the user types while on the other are a list of relevant attributes. How each category of users is derived will depend on the specific design project. Ideally, the user types will emerge from user research. This may either be from existing secondary data (e.g. census data, reports, etc.) or primary data, such as questionnaires (p.102) or interviews (p.78). Each profile should convey relevant trends found for each chosen attribute. This may include demographic data, occupation, education level, income, skills, technology expertise, attitudes, and so on. Viewed side by side within a table, user profiles can help the designer visualise comparative trends. Viewed singularly, an individual user profile can form the basis for creating a persona (p.100).

While personas provide a narrative description of a typical user, user profiles present information as a range or trend. For example, a user profile may be labelled as a 'Tech savvy super-user', within the range of 19 to 30 years old, on average 70 percent female and working 40 to 55 hours per week. While a persona might be emblematic of that profile type, personas and user profiles do not follow identical formats. User profiles are not static but subject to refinement through each iterative stage of the design process.

EXERCISE

YOU WILL NEED
A partner, pen, paper, internet

In this exercise, you will create user profiles for a product or service. Focus on your own design problem, or follow the 'Designing Space Travel' brief (p.141). Use existing data from the resources on the companion website or generate your own. Use the provided template (p.194) to document the results.

1 **Choose a target user population.** Use one of the following:
- Your classmates/colleagues, or
- Your social network.

If you are using the existing data from the resources on the companion website, you can skip this step.

2 **Find out information about your user population**, using a variety of methods. Survey, count, observe, perform online ethnography or do mini interviews. Data may include:
- Demographic data, skills, education, occupation
- Information specific to the design problem.

E.g. *travel frequency, country of origin, preferred airliner, reason for travel*

If you are using the existing data from the resources on the companion website, you can skip this step.

[20 minutes]

3 **Look for trends in key attributes.** Look for attributes that unite some users and distinguish them from others. What these categories are and how you form them depends on your data.

E.g. *reason of travel: fun, work, visiting family*

E.g. *planning behaviour: plans ahead, books last minute, joins with friends*

What types of user can you see emerging in relation to the design focus?

[10 minutes]

4 **Identify the most recurrent types of users**, and add these to the user profile template (p.194). Add any important attributes you have identified into the empty rows of the table.

[5–10 minutes]

5 **Check that you have an adequate range of user profiles** represented in the table and that they are distinct enough from each other. Give each user type a nickname.

E.g. *'Fun on a budget' or 'Luxury holiday-maker'*

[5 minutes]

6 **Get feedback from another group and discuss.** Review each other's user profile tables. How would your design meet the needs of the different user groups. Is there an opportunity to address a marginal user group? Make revisions, if required.

[5 minutes]

Design. Think. Make. Break. Repeat. **131**

Value Proposition Canvas

Addressing customer pains and gains

ACADEMIC RESOURCES:

Christensen, C. M., Anthony, S. D., Berstell, G., & Nitterhouse, D. (2007). Finding the right job for your product. MIT Sloan Management Review, 48(3), 38.

Johnson, M. W., Christensen, C. M., & Kagermann, H. (2008). Reinventing your business model. Harvard business review, 86(12), 57-68.

Osterwalder, A., Pigneur, Y., Bernarda, G., & Smith, A. (2014). Value proposition design: How to create products and services customers want. John Wiley & Sons.

Designing products and services that offer pleasurable experiences can provide a strategic advantage and help solutions to stand out in a competitive landscape. For example, when ride-hailing company Uber entered the market, it offered passengers a more pleasurable experience compared to that of the often unreliable taxi services. Uber managed to resolve many of the pain points existing with taxi services at the time while providing an overall better experience to passengers, such as real-time tracking of vehicles and mobile payment of ride fees.

To design products or services that provide a pleasurable experience, it is necessary to understand the 'value proposition' that a solution offers – the underlying reason a customer would want to engage with the solution. The value proposition canvas method builds on the business model canvas (p.30) to guide the process of designing a solution that addresses customers' pain points (their issues, annoyances) and gains (what they want to achieve). The method starts with selecting a customer segment or user group and understanding their objective – what task are they trying to get done, what are they trying to achieve, how can the solution be designed to assist with this task?

The method is commonly used to explore new customer segments, that have not previously been considered, and to determine whether a particular bundle of products or services can address their needs. It helps indicate any of the customer's unaddressed pain points, as well as determine whether future opportunities to create value for that particular customer segment have been missed. The method can also be used to focus on customer needs for a customer segment that has already been considered in a solution, and ensure their needs are adequately addressed.

EXERCISE

YOU WILL NEED
Pen

In this exercise, you will fill a value proposition canvas for a target customer of your choice, using the template provided (p.189). Focus on your own design problem, or choose a design brief (p.138).

1 **Choose your target customer** (you could use a persona, p.100) and keep them in mind when completing the circular part of the template. Write down a two to three word description of your target customer in the 'customer segment' title block.
[5 minutes]

2 **Question what this customer is trying to accomplish** and jot these ideas down in the 'customer jobs' section of the template.
[10 minutes]

3 **Question what annoys the customer, before, during and after getting their jobs done.** Note ideas down in the 'pains' section of the template.
[10 minutes]

4 **Question what outcomes and benefits the customer wants.** Write these down in the 'gains' section of the template.
[10 minutes]

5 Move to the square part of the template and **ask what products or services exist that enable the target customer do the jobs** you have listed? Write these down in the 'product and service' section of the template.
[10 minutes]

6 **Reflect on how the products and services you listed overcome the pains** you identified in step three. Write these down in the 'pain relievers' section of the template.
[10 minutes]

7 **Reflect on how the products and services you listed create customer gains** by giving customers what they want. Note down ideas in the 'gain creators' section of the template.
[10 minutes]

8 Review the whole template and **explore whether all pains and gains are being addressed by the pain relievers and gain creators** by drawing lines between possible combinations. If anything is left unlinked, think of how these could be addressed with a new design.
[20 minutes]

Video Prototyping

Communicating design concepts through video narratives

ACADEMIC RESOURCES:

Mackay, W. E., & Fayard, A. L. (1999). Video brainstorming and prototyping: techniques for participatory design. In CHI'99 extended abstracts on Human factors in computing systems (pp. 118-119). ACM.

Markopoulos, P. (2016). Using video for early interaction design. In Collaboration in Creative Design (pp. 271-293). Springer.

Tognazzini, B. (1994). The "Starfire" video prototype project: a case history. In Proceedings of the SIGCHI conference on Human factors in computing systems (pp. 99-105). ACM.

Vertelney, L. (1989). Using video to prototype user interfaces (pp. 57-61). ACM.

Video prototyping is a useful method to communicate an idea before building functional prototypes. A video prototype is a short movie showing how one or more users would interact with a future product based on a scenario (Markopoulos, 2016). This allows rapid recording of concepts in an engaging visually-rich format. Using videos is beneficial for providing the context of use, communicating a narrative around experiencing a product or service, and showing interactions between people and their environment including facial expressions and body gestures.

Video prototypes can take various levels of fidelity depending on the project stage. At an early stage, and for communication inside the team, rough videos are sufficient as the focus is on exploring ideas for concepts. At such an early stage, it might be useful to create a video prototype in combination with experience prototyping (p.58) and role-playing (p.108), as this allows quick play through and iteration on interaction scenarios with valuable contextual information.

More advanced video production techniques may become necessary when the concept is further refined or shared with a wider team that may include project stakeholders. The more polished the video prototype, the more convincing the depicted scenario will be. Using mock-ups (p.90) in combination with other techniques can create the impression of a fully working product or service. Useful techniques for creating video prototypes are varying shots, shooting a sequence, stop motion animation, 'Wizard-of-Oz prototyping' and green screen keying.

Video prototypes can be used at any stage of a design process to explore concepts and their implementation, to test concepts with prospective users, to communicate concepts to an external audience and to pitch concepts for funding.

EXERCISE

YOU WILL NEED
Smartphone with camera, paper, cardboard, masking tape, Post-it notes, pens, video editing software (optional)

In this exercise, you will create a 30-second video prototype to convey an idea for a product or service. You will learn how to use techniques to represent the interactions between people and a product or service through video.

1. **Select a problem or situation** to redesign.
E.g. how can you improve the experience of finding a room in building X?

2. **Brainstorm a solution** for the chosen problem or situation. You can also use a solution from a previously completed exercise, such as an experience prototype (p.58).
[10 minutes]

3. **Write a usage scenario**, in which a user interacts with your new design concept. **Identify which interactions** you want to convey in your video prototype. You may need to prioritise certain parts of the experience to ensure they fit into a 30-second narrative.
[10 minutes]

4. **Decide which shots you need** to tell the story of the experience and create a storyboard using the provided template (p.187).
[10 minutes]

5. Based on the storyboard, **identify the locations, actors and materials** you will need to film the video prototype. You can use previously created design artefacts, such as experience prototypes, in your video, or create new artefacts to use in the video.
[10–30 minutes]

6. **Shoot your video**. Before the actual shooting, rehearse the full scenario. Use your phone to record the video.
[20–30 minutes]

7. Optional: If you have access to video editing software, **edit the video snippets** into a final narrative. You can also add text title slides or subtitles to explain the interactions. Add sound and voice-overs that match the context and scenario.
[30 minutes]

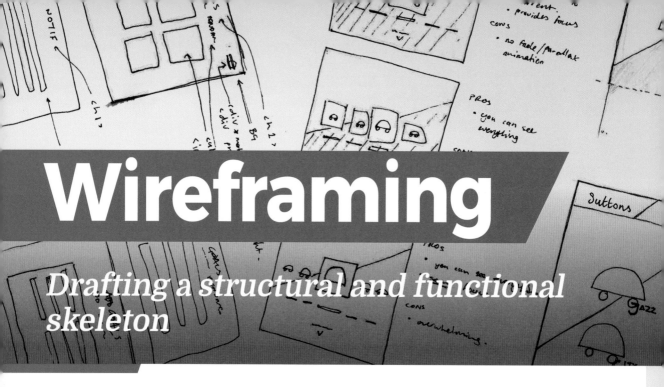

Wireframing

Drafting a structural and functional skeleton

ACADEMIC RESOURCES:

Garrett, J. J. (2003). The elements of user experience: user-centred design for the web. New York: American Institute for Graphic Arts.

Wireframes are line drawings that show the fundamental structure or functions of a product or system. As an extension of sketching (p.116), wireframes offer a formal way for designers to think and communicate what a design can do. They do this without surplus high-fidelity information, such as images, fonts, colours or typography, that may distract from the core purpose. Because wireframes do not look or act like a finished product, they are quick to produce and invite modification.

Wireframing has long been a core element of 3D modelling and product design and is now also central to the design of digital applications. Conducted early in the ideation process, wireframes, like low-fidelity prototypes (p.84) and mock-ups (p.90), allow us to shift from researching and understanding a problem to building a solution. They can be especially useful for expanding upon conceptual models, for example, those that are generated through card sorting (p.34). In this way, the users' mental models and early sketches can be combined and translated into structures, such as menus, tabs, headings, and pages.

Just as architectural blueprints have their own codified visual language, wireframes give us the symbols to express, reflect on and communicate our ideas for how a system will work. They can either be built on paper or using illustration software applications (e.g. Adobe Illustrator) or specialised software tools (e.g. Balsamiq). To streamline workflows, some software tools even allow for the addition of simple interactivity to simulate the transition from one screen to another. Wireframes aren't very effective when testing with users since most users won't be familiar with the meaning of the codified symbols used in wireframes. However, wireframes can be used to seek feedback from other designers or domain experts before turning them into mock-ups or prototypes.

EXERCISE

YOU WILL NEED
A partner, pen, paper

In this exercise, you will create a set of wireframes to represent the user interface of a mobile app or website solution. Choose a design brief (p.138), such as 'Designing Space Travel' (p.141).

1. **Brainstorm a solution** for the chosen problem or situation. You can also use a solution from a previously completed exercise.
[10 minutes]

2. **Sketch the parts of your product** that you want to represent in your design. For example, if you are following the 'Designing for Space Travel' brief you can choose a specific scenario
E.g. finding the cheapest flight to a certain destination within two weeks
[15 minutes]

3. **Prepare a canvas** for every screen that you need – this is a rectangle with the same proportions as the screen's display format. For a search function, you are likely to have a 'home screen' and 'results screen' at a minimum. You can use the provided template (p.195) for mobile apps. For desktop browsers, use the provided resource on the companion website and draw your own canvases.
[5 minutes]

4. **Draft up your navigation and functions for each screen**. Use the suggested icons provided in the resource on the companion website and draw your own if needed. Wireframes don't need to be pretty; quick line drawings that show careful thought about the product's functionality are better than perfectly drawn illustrations. Use existing principles, such as patterns or *Gestalt*, and your experience of common web and mobile app functions to create a solution that users will understand.
[40 minutes]

5. **Present your screens in a linear format** so that they can be viewed in sequence. **Add annotations** to explain important functions. You can use colour to highlight the key features of each wireframe, but it should not be used as a visual design element.
[10 minutes]

6. **Discuss and evaluate your wireframes** with your partner. Make sure the wireframes makes sense to others and modify them based on feedback.
[15 minutes]

Design. Think. Make. Break. Repeat.

Design. Think. Make. Break. Repeat.

Design Briefs

Autonomous Vehicles

Self-driving cars are becoming a reality due to advancements in machine learning and AI. They are robots disguised as cars, using sensors and actuators to navigate roads and respond to features in the environment, and they are joined by many other robotic systems that are lining up to become part of the fleet of autonomous systems in future cities.

Today, Google has several driver-less cars that have proved capable of manoeuvring the roads. Likewise, in 2014, Telsa rolled out a new autopilot feature. The feature was released in the form of a software update: no new car required. Although completely self-driving cars are estimated to still be some years away, these first attempts at mass-producing autonomous vehicles for today's cities demonstrate that as designers we need to start thinking about how people in the future will interact with autonomous vehicles.

It may be machine learning and AI that are bringing driverless cars into a close reach of the mass consumer market, but it will be the user interface and user experience issues that will make or break the future of autonomous vehicles.

This design brief focuses on screen-based interfaces that enable interactions with autonomous vehicles. Your task as a designer is to map out what will be required to assist people in interacting with autonomous vehicles in future cities.

The futuristic nature of the design brief will make it difficult to gather first-hand experience of autonomous vehicles and their usability issues. You may have to base some of your research on pre-existing documentation and research found online, such as driving tests of autonomous vehicles or academic research papers.

You should identify either a problem scenario (e.g. garbage collection in cities) or a type of autonomous vehicle (e.g. delivery drones). This focus might also emerge while doing background research on the design problem, which should involve opportunity-seeking. You are not required nor expected to prototype any of the physical aspects of your chosen autonomous vehicle platform. Instead, you are asked to design the visual interface that allows people to interact with the autonomous vehicle. For example, this could be in the form of:

- a mobile app used to book shared autonomous vehicles,
- a mobile app to order and track products delivered by urban drones,
- a web application to control a fleet of garbage collection robots, street cleaning robots or air quality measurement drones, or
- a visual control interface integrated into a driver-less car.

You should identify a specific use case and focus on designing a solution for that specific use case, narrowing down the solution. More important than covering all aspects of your chosen platform, it will be to thoroughly think through the interactions for that specific use case and to use iteration and evaluation to get the details of the design right.

Designing Space Travel

People have dreamt of travelling through space for centuries. In this design brief, you will get to imagine what form of interactive applications might be necessary to book space travel in the future. In 2016, for the first time, a rocket launched into space and safely returned to earth. Elon Musk, founder of SpaceX, is planning to combine this super rocket with a spaceship built to carry at least 100 people – coined as the first Interplanetary Transport System (ITS). SpaceX plans to land an uncrewed capsule on Mars in 2018. According to Elon Musk, we will be able to establish the first human colonies on Mars in 50 to 100 years. The ITS will be able to make the trip in about 80 days, depending on the position of Earth and Mars at the time of travel. Once the first colonies are established, it won't be long for the tourism industry to discover space travel to Mars. Their target audience could reach from anyone wanting to visit their relatives living on Mars to people taking time out from their job or life for an extended holiday with great views of Earth along the way. Only, how are they going to book their trip to Mars?

Your task is to design an online booking interface for [insert a funky name of your choice here, e.g. "SpaceJet"], a new aggregator service for booking flights to Mars. Their catalogue includes one-way flights for colonisers, return flights for visitors, and holiday packages for vacationers. The booking interface should be available to customers using either a browser on a laptop or smartphone, such as an iPhone or Android phone – assuming those devices will still be around in 50 to 100 years. For the purpose of this project, you can also assume that people will still interact with laptops using a keyboard and mouse pointer, and with smartphones through multi-touch screens.

You may also want to consider additional aspects in your solution, such as:

- Thinking about the whole experience; not every city will have a space station (even in 100 years), people may need to board a connecting flight from their city to arrive at the nearest space station.
- Even in 100+ years a trip to Mars will still take several months; people may therefore want to book more than their meal choice. Think about how the trip can become a destination itself and how to allow people to choose those experiences as part of the flight booking process.
- There may be already a lot of competition offering similar services; how can you set your booking interface apart from others. For example, think about additional experiences you could offer on Mars as part of a holiday package.
- Think about the experience on Mars itself and how to enable users of your booking interface to choose aspects of that experience beyond hotels, such as rovers to get around or the style and type of space suit.
- There will be more than one spaceline in addition to SpaceX; think about how those could be represented in your booking interface.

You are free to make up details required for designing your booking interface, such as the price of a flight to Mars, but try to keep things realistic – for example, the price of a flight to Mars might be roughly 100 times the costs of an international flight on Earth.

Museum Visitor Experience

While some people can spend the entire day exploring museums, others would not mind if they never set foot inside one again. This is a potential problem for an institution that relies on public interest for funding: museums need to be interesting and relevant for everyone. The most critical issue facing many museums today is how to remain relevant, and engage broader audiences in new ways.

For an industry that has traditionally defined itself in being an authority on cultural matters and the wardens of historical collections, the rise of the information age has changed the game. Museums are no longer the gatekeepers of knowledge that they once were, and much of their content can be accessed from the comfort of your sofa. Even artworks have become digital goods. If you want, you can view the Mona Lisa from the comfort of your home, download it, print it and hang it on the wall. Museums are in the midst of a digital disruption and are constantly challenged to reinvent themselves.

Spurred on by this relevance crisis, many museums have been busy adapting their practices. To maintain their unique position in the community, they need to offer unique, immersive and astounding visitor experiences that bring people to the building. And you can help them do it! For this design brief, your task will be to select a museum of your choice and to design an experience that builds upon the collection that the museum offers. It should improve visitor engagement, fulfil educational goals of the institution, and help the museum to attract new audiences (or affirm their relationship with existing ones). The experience should be interwoven with the building itself; it could not exist without the collection and the environment of the museum. The experience is as broad or as narrow as you decide – it could begin at the door of the museum or focus on a particular gallery.

Using the methods and templates in this book, you should explore a variety of different aspects:

- Tangible: making use of the physical qualities of the artefacts and collection
- Spatial: using the building itself and its spatial qualities as a setting for the experience
- Narrative: telling a story using the collection of the museum
- Digital: using technology to support new ways of engaging and teaching visitors

Design an experience that the visitors will never forget..

Supermarket of the Future

Almost everyone has, at one point in time, experienced the act of grocery shopping. Some may even describe this activity as a weekly chore. This design brief is set within the context of grocery stores and the broader experience of and around shopping, both from a customer as well as provider perspective.

Grocery stores exist in various sizes and locations, and serve a diverse range of customers and their needs. They can reach from small 'box stop' grocery stores at a petrol station to massive supermarket halls inside shopping centres, each with their own unique context, challenges and opportunities.

Your task is to develop a vision for future shopping environments (both digitally and physically) that improve the experience of customers and staff. To achieve this, you will need to first understand the context, define your problem area, and understand the various stakeholders involved. Solutions should consider the interests and perspectives of different stakeholder and user groups, which can be of conflicting nature. Think, for example, about the shopping requirements for an elderly, retired person versus a young full-time professional on the way to work, and the social and behavioural implications of these scenarios.

You should use the methods and templates from within this book to scaffold your design process. It will be your task to first identify a problem area to work on, understand this problem space, explore pain points of current experiences – either of customers, staff (such as shop assistants) and any other stakeholder contributing to the shopping experience.

Based on your specific problem context and stakeholder group, you will then ideate solutions that address all or some of the identified pain points of this problem.

You should then employ various prototyping techniques, with the aim to produce as many prototypes as quickly as possible. By prototyping ideas as tangible manifestations that can be experienced you can identify and refine solutions that best meet the needs of the stakeholder.

This brief challenges you to develop solutions that address the problem from a human-centric perspective, not in a technology or materials-driven approach

Design. Think. Make. Break. Repeat.

Case Studies

Autonomous Vehicles

Questionnaire & Interview Results

Participants' conditions:

60% With a physical disability
20% With visual difficulties
20% With long-term illness
20% With mental health difficulty
6.7% With hearing loss
6.7% With speech difficulty

Current modes of transport:

36.4% Public transport
22.7% Car
18.2% Private
9.1% Taxi
9.1% Friends or Family
4.5% Walking / own medium

Applications used for navigation:

Most used: Google Maps
Second most used: Uber
Third most used: TripView real-time transport app

Key comments:

"I have difficulty driving, I try to do it for limited duration or contact my closest relative to drive me around."

"I need assistance to get on the train or vehicle mostly always."

"Being in a wheelchair can make transportation anywhere very difficult. Public transport needs to be wheelchair accessible, which sometimes limits the amount of services I can use as some older vehicles are not wheelchair friendly."

"As I am unable to drive myself, travelling long distance means I need to find somebody who is willing to drive me (generally mum or dad) to where I need to be."

Interview Results

Purpose: Collect data on key questions and narrative stories about people's daily scenarios.
Method: Semi-structured interviews with five participants with physical disabilities

Where participants need to travel:

Work, educational institutions, recreational activities

Most frequent type of transportation:

Taxis, family car, private organisation from the community group

Reasons why people don't use public transport:

Buses are not pleasant or unreliable.
Assistance on trains is set-up to only help one person at a time. If there is more than one person with physical disability trying to get on the train, assistants lose control of time and there are not enough ramps to provide access.
Waiting time is too long. Not all areas around Sydney make it easy to access vehicles.

Key insights about pain points:

Almost impossible for people with certain physical disabilities to move around the city independently
Lack of easily accessible transport options
Unhelpful attitude from drivers and passengers
Accessible vehicles are rare and expensive.

Storyboard

This response to the Autonomous Vechicles brief specifically focused on people with disabilities. The response employed various design methods to identify their current pain points and to devise solutions.

Persona

Age 23
Occupation Student
Status Engaged
Location Lilyfield
Physical Disability Cerebral Palsy

Bio

Richard has been using a wheelchair since birth. Now that he is a university student in his second year and regularly takes the bus from home to go to class. There is only one bus route that he can take and the buses arrival time can be very unpredictable.

He often has to rush to his lecture, which makes him very physically tired and mentally frustrated.

The weather can make travels difficult for him as the bus stop near his house is uncovered. For which issue, he opts to try and book a wheelchair accessible taxi, hoping there is one available for him so he can get to university on time.

Motivations

Richard Loves to Study and he has dreams as many other people.

He likes using technology as long as it is appropriate to his condition.

Goals

- He hopes to see the public transportation to improve and provide more access to people with disabilities
- He wishes that in the future there will be a more reliable service or other more accessible forms of transportation that satisfies his needs.

Frustrations

- He can't make use of the bus stop next to his house on bad weather.
- Getting tired and mentally frustrated due to the unpredictability transport services can be.
- Long waits for private taxis or taking long to get to the station or bus stop.

Site Map

Home

Dashboard

Map

Destination

Driving

GPS View

Parking

Calendar

Detailed Event

Air Options

General Overview

Design. Think. Make. Break. Repeat.

Low-fidelity Prototype

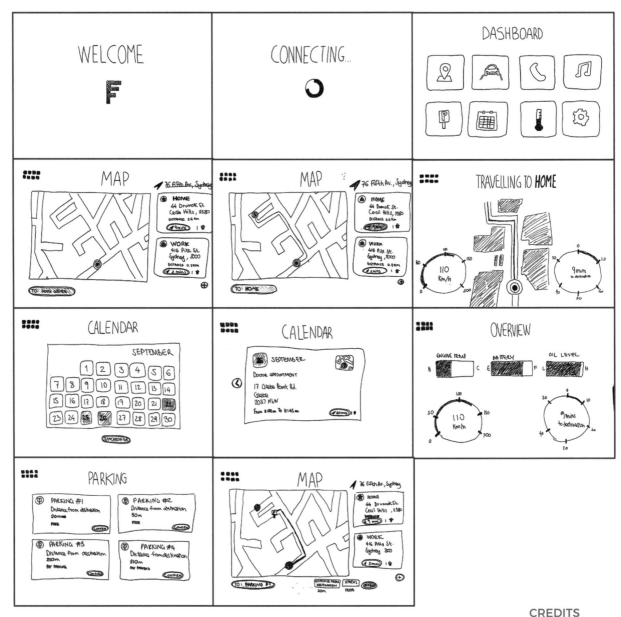

CREDITS
Francesca Serpollini
Francisco Acosta

Case Study – Autonomous Vehicles

Usability Evaluation

Mock-ups

Designing Space Travel

Online Ethnography

	Webjet iOS	Flight Center Android	Expedia iOS	Skyscanner iOS
Total Reviews	823	14	1808	33568 (but only 10,000 reviews analysed)
Topics v Star Rating >	2.5	3.9 Stars	3.5 Stars	4.5 Stars
Bugs				
Sentiment (out of 10)	0	na	2	2
% of total reviews	56.2%	na	8.5%	2.2%
Trending	decreasing mention	na	low and steady	decreasing mention of bugs
Comments		na		
Design & UX				
Sentiment (out of 10)	8	10	8	7.5
% of total reviews	28.7	26.6	19.7	19.8%
Trending	increasing mention	not enough data	steady increase then sudden drop in mention of praise	Decreasing mention of design
Comments	"Great" and "Easy" appeared in 36% and 68% of overall reviews respectively	See summary below	Many comments praising ease of use, and speed of use "easy" - 40.3% of reviews. "Book" however had medium sentiment (5/10) and appeared in 29.7% of reviews.	"Easy" and "Simple" dominated this category appearing in 38% of reviews!
Satisfied Users				
Sentiment (out of 10) will always be 10 here	Always 10	10	Always 10	Always 10
% of Matches	16.2	14.3	23.7%	31.9%
Trending	increasing sharply recently	not enough data	steady increase then sudden drop in mention of praise	steady
Comments		See summary below		Popular words include: "Great" (2643 mentions accounting for 26.4% of reviews!) and "Love" (1514 mentions accounting for 15.1% of reviews!) Additionally, "Helpful" rated highly with positive sentiment at 744 mentions

UX sentiment was mixed, 50% of users reviewing UX were positive
- Since **June 2014**, UX sentiment has trended **positive** - "Great" and "Easy" appeared in 36% and 68% of overall reviews respectively

"Very simple and efficient to use." - Cellisootyashr[1]

"Very easy to use and handy to book flights quickly" - Raeange[1]

"Pretty easy to navigate through site" - Sah! Steve[1]

aggregation and **filtering** are popular

"I found this app extremely useful when I was interstate for work and couldn't remember my flight details - simply logged in and there they were!" - Paul Pichugin1341[1]

"Comparisons are simple and it takes away the time spent going into individual apps for plane carriers. Convenient, reliable and simple to use." - JFH1967[1]

Unclear fee structure

Of all **4001** UX related reviews, around **70%** had **positive sentiment.**

Typical positive UX reviews include mention of being quick, easy and logical to navigate, some users reporting clarity provided by colour coding

"I have used this app to book my last 3 holidays. The best feature for me is the price alerts so I know if the prices are changing if I am not quite ready to book yet. Easy to use, and I will be using it to book my next flights too!" - Diba03[2]

 1-2 Star reviews
Significant insight: Some users **don't care** exactly **where** or **when** they go, they just want the **cheapest** of flights

"Used to be good When I originally used this app, 3 or 4 years ago it was great, <snip> I'd choose destinations based on flight price, so went to some unexpected places, great. Alternatively if you knew where you wanted to go your results could tell you when in the year the prices where cheapest. Now you have to know departure airport, destination & dates, what's the point of that, I can search google for that. It's not letting me find cheapest times to travel." - Spitfirenellie[2]

ETHNOGRAPHY RESULTS (summary)

For full details on ethnography and **sentiment analysis** please see appendix section.

This response to the 'Designing Space Travel' brief involves data collection through online ethnography and interviews, synthesised through affinity diagramming and turned into a prototype for usability testing.

Affinity Diagramming

Peace of mind for medical issues

- I want to easily match my insurance cover to my travel itinerary

Personal items are important on long trips

- I want to know what I can bring with me
- I want my cabin to feel like home away from home

An easy booking process

- I want to book tours easily
- I want to be able to easily choose my room

- I want a trip of a lifetime
- I want an experience based on my preferences

Personalised Experience is important

- I want to be inspired with recommended experiences

Lifestyle Interests

Health and Medication	Photography	Souvenir	Fashion	Friendships	Technology	Social Status

Design. Think. Make. Break. Repeat. 151

Usability Testing

The think aloud protocol described in (Lewis, 1982) was used in an initial round of user testing for key user tasks. The participate was mid twenties, male and a peer expert in user interface design principles, making him an ideal candidate for round one testing. Several factors were observed for and recorded including: body language, hand gestures, other non-verbal cues, facial expressions, fluctuations in speech and tone of voice. These interactions were captured with the behaviour forms (see Appendix B).

We found both video recording and screen recording with audio to be invaluable in the recording process. We analysed the results using the affinity mapping method from (Hanington & Martin, 2012) to draw out usability themes based on the 5e's (Quesenbery, 2002) with additional notes as to why the issue was categorised as such ie; lack of clarity, wrong context, etc. This enabled patterns to be established and identification of problems consistent across all three interfaces. The results of the usability testing along with an update to the information architecture using a card sort process from (Rugg & McGeorge, 1997) were incorporated into wireframes.

A final round of user testing was conducted using the think aloud protocol and collected additional data using the System usability scale (Brooke, 1996), and non-verbal behaviour (University, 2016) to highlight further issues with the design (see Appendix B).

This round was conducted with 3 additional participants from a wide range of backgrounds.

"I'm not sure what I'm scheduling - what is the start and end date for??"

"these I assume are things I assume I saw on the inspire page that I can compare price? WRONG!!! It's comparing itineraries!"

REFERENCES

Brooke, J. (1996). SUS-A quick and dirty usability scale. Usability evaluation in industry, 189(194), 4-7.

Hanington, B., & Martin, B. (2012). Affinity Diagramming. Universal methods of design: 100 ways to research complex problems, develop innovative ideas, and design effective solutions, 12-13.

Lewis, C. (1982). Using the" thinking-aloud" method in cognitive interface design. IBM TJ Watson Research Center.

Quesenbery, W. (2003). The five dimensions of usability (Vol. 20, pp. 89-90). Mahwah, NJ: Lawrence Erlbaum Associates.

Rugg, G., & McGeorge, P. (1997). The sorting techniques: a tutorial paper on card sorts, picture sorts and item sorts. Expert Systems, 14(2), 80-93.

CREDITS

Iris Aranguren
Aaron Blishen
Benjamin Marell

Usability Test Results

Platform	Welcome Page	Explore/Inspire	Plan/Scrapbook	Create Itinerary
DESKTOP		-Two search options -Not clear what groups are when adding to plan -The group by menu is not clear -Inspire word is not clear to meaning	-Not clear as to how to group activities -The search results on scrapbook clear as what they are for -Add to scrapbook affordance visibility	-The compare icon is not clear -The compare icon is in the wrong context -Recent activities is not consistent across activities
iPAD	Tester wasn't sure of difference between inspire and quicksearch	-Not clear that adding to scrapbook is a recommended action -Appreciated filtering options, thought that was clear -Would appreciate if it was clear which items were favourited	-Search results clarity - what am I searching? -Not obvious results can be added to scrapbook or how it is to be used, no affordance for drag and drop and not obvious you can create / move groups.	-Unclear of context of date selector - what are dates being selected for? -Unclear of purposes of suggested itinerary page is -No back button in customise itinerary
ANDROID	-Sign In: obvious button, good, big -GO button, Log In or Sign In would be better. More clear.	-Not sure of purposes of this screen -Favourite icon confusing, is it used for liking?	-Interface too busy -Favourites heart button not clear in this case, plus button better -How do I add more activities?	
COMMON THEMES AND KEY FINDINGS	Needs to be clear from the outset what the user can do - quickly search for flights, or be start planning a journey. Login needs to be clear.	-The fundamental concept of saving activities away into a scrapbook **was poorly explained and not understood** by participants. -Additionally, once this was explained to testers, it was still not clear how to add such items to the scrapbook, due to poor iconography choice (we used a heart icon).	-Like the Inspire page, it was not clear what the scrapbook was for. It was also not clear that the inspire page was connected to the scrapbook. -The context of the search on the scrapbook page was not clear. It wasn't clear that the search bar results could be added to the scrapbook.	Again, it was not clear what was happening. The user thought they were already organising their journey in the scrapbook, and then were confused when they were presented with itinerary options, and then the opportunity to customise them. **The flow was unclear.**
SOLUTIONS	Explanatory text for the different options Login Clarification	After card sorting, "Inspire" was changed to "Explore" to be more user-goal focused. Iconography was changed, and on larger form-factor platforms the text "Add to Plan" was added.	After card sorting, "Scrapbook" was changed to "Plan" to be more user-goal focused. An overlay tutorial to demonstrate that activities can be searched and dragged and dropped (see Appendix) "Add to Plan" buttons were added on each activity.	After card sorting, "Schedule" was changed to "Itinerary" to be more user-goal focused. Similarly, the above flow issues were solved with wording/labelling changes that linked screens. For example from the Explore Page "Add to Plan" replaced the favourite button., On the Plan page > "Suggest Itineraries", was added, and then "Customise your itinerary" was added to the Itinerary page.

High-fidelity Prototype

Museum Visitor Experience

Research Visualisation

UNDERSTANDING *The* VISITOR EXPERIENCE
ENGAGING YOUNG ADULTS IN ART GALLERIES

Young adults don't feel as if art galleries are for them. There is disconnect between the culture of galleries and the culture and identity of young people.*

SO HOW DOES AN ART GALLERY ENGAGE A YOUNG AUDIENCE?

A SOCIAL ATMOSPHERE
YOUNG PEOPLE ARE HIGHLY INFLUENCED BY THEIR PEERS, THEY SEEK OUT COMMUNITY AND GRASSROOTS EVENTS WHERE THEY SHARE A COMMON INTEREST.

Creating a social destination is essential in attracting young adults to art galleries. Providing an atmosphere where young people feel comfortable is important. Young people also desire an environment where they can socialise with like minded people, creating a sense of belonging or identifying to a movement.

" I love finding new music, the whole discovery process, I just need to find the next thing that tickles my fancy "

50%+ DISAGREE
MUSEUMS ARE FUN & SOCIABLE PLACES FOR YOUNG ADULTS
40% - AGREE
5% - STRONGLY AGREE
5% - STRONGLY DISAGREE

NEW EXPERIENCES
DISCOVERING NEW EXPERIENCES MOTIVATES YOUNG PEOPLES TO IMMERSE THEMSELVES IN A SUBJECT.

Young people will seek out unique and unexpected experiences that are culturally relevant to them, finding inspiration and entertainment in the process. The excitement of discovering something new motivates young people to immerse themselves in a subject.

" I hate large groups of people; parents screaming at the kids, the people there are annoying! "

ADDING CONTEXT
PROVIDING ADDITIONAL LAYERS OF INFORMATION SURROUNDING ART PROVIDES VISITORS WITH A GREATER APPRECIATION OF THE WORK.

Young people expressed that viewing art knowing its broader context creates a more enriching experience. Having a deeper understanding of the history surrounding the art allows for a greater appreciation of the piece. They expressed a desire to want to ask questions to find out more information on how the piece was created and the background of the artist.

MOTIVATIONS FOR VISITING A MUSEUM 17.5%
A SOCIAL OUTING WITH FRIENDS

INSPIRATIONAL LEARNING
INTERACTIVE LEARNING EXPERIENCES ARE EXPECTED TO MAINTAIN INTEREST. THEY MUST BE CHALLENGING AND PROVIDE A FORM OF LEARNING PROGRESS.

With the availability of the internet, young adults expect to interact in a dynamic learning experience. Visual learning aids enrich the learning process and creates engagement. Young adults need to be challenged by the learning experience and understand that persistence is necessary to progress.

INFORMATION DISCOVERY
PERSONAL EXPLORATION NEEDS TO BE ENCOURAGED WITHIN A GALLERY. YOUNG ADULTS LIKE TO DISCOVER INFORMATION AT THEIR OWN PACE.

Young adults want and mix of freedom and direction within a gallery. They desire some semi-guided experiences, but also want the ability to customize it by not be locked into something if it doesn't interest them.

Francesca Serpelloni   Mackenzie Etherington
Alex Elton-Pym   Matt Fehlberg

CREDITS
Francesca Serpelloni
Alex Elton-Pym
Mackenzie Etherington
Matt Fehlberg

CREDITS
Ethan
Daya

154 Case Study – Museum Visitor Experience

This case study focuses on making the visitor experience of a museum more enjoyable for millennials through the use of personalisation, interactive technologies and context-aware information to arrive at a friendly social outing for groups.

Storyboards

Jane likes History and decides to visit the Museum for a fun day out by herself.

She buys a ticket and on entry to the museum itself has her ticket exchanged for what seems to be a hologram.

Jane is intrigued and pleasantly suprised that the hologram is a historical assistant and guide.

She tells the assistant about her interests and it replies directing her to the exhibits which suit her best.

As she wanders through the different collections Jane asks Alex about pieces around her and what might interest her.

Jane is hungry and asks "Alex" where to eat. Alex begins navigation to a restroom with the cafe as the next stop.

Jane is now leaving, but before she gives Alex back she connects her phone to Alex to save the exhibits for later.

Jane leaves and shares her experience with friends, encouraging them to get their own historical assistant too.

John and Amy are on a date. They want to do something different; they decide to visit the Museum.

While they are browsing through the Egyptian Exhibition an artefact speaks to them, taking them by suprise.

Amy spots the "hold to talk" button next to an artefacts. She presses the button and asks a question.

To Amy's delight the artefact responds. John's interest is now piqued and he moves in closer.

John is intrigued and wonders how interactive the exhibit is, so he asks a random but related question.

The artefact answers, and also provides additional information about the other talking pieces in the collection.

Now interested in the talking artefacts they follow the directions to the next part of the collection to learn more.

They arrive at the artefact and are excited to find another part of Egyptian history to talk too and learn more.

Design. Think. Make. Break. Repeat.

Video Prototype

CREDITS
Alex Elton-Pym

High-fidelity Prototype

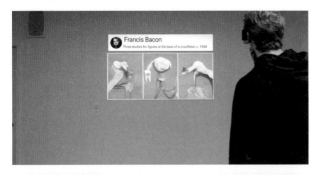
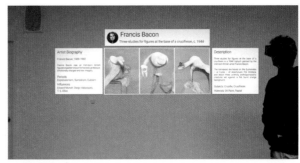
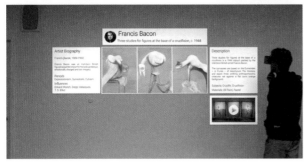
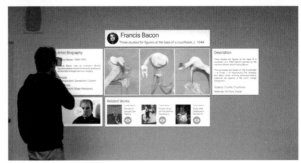

CREDITS
Mackenzie Etherington

Decision Matrix

Criteria		AI Companion	Hologram Assistant	Living Pieces	Talking Exhibits
Physical Criteria					
Tangible Interaction		2	3	3	3
Spatial Considerations		1	1	3	2
Interactivity		3	3	2	3
Implementation of Tech		3	3	2	2
Scale		3	2	3	1
Aesthetic		1	1	3	-1
Usability Criteria					
Self - Navigating		2	3	0	1
Error Tolerance		-3	0	2	1
Age - Agnostic		1	1	2	2
Insightful Information		3	2	3	2
Understandable Info		2	3	3	1
Fun Factor		2	2	2	3
Engagement		3	2	3	3

CREDITS
Ethan Daya
James Horne

Supermarket of the Future

Competitor Analysis

METHODS

CONTEXTUAL MAPPING
Conducted at mid-day in Wynyard Station Coles Express
Monday at 12:30pm - 12:45pm

CONTEXTUAL OBSERVATION
Conducted at mid-day in Wynyard Station Coles Express
Monday at 12:45pm - 1:00pm.

Observations
- More customers were in groups of 2 than in Woolworths Metro, possibly as a result of the location inside a train station.
- There were less office workers than in Woolworths, this was probably as it was during office hours and most workers would not be travelling on public transport.

Key Insights
- The placement of the store changed the customer based significantly, with Coles having less officeworks and more small groups than Woolworths
- Having large shelving made navigation for customers difficult as signage and visual paths were obscured. This caused shopping to take longer as customers walked a lot slower

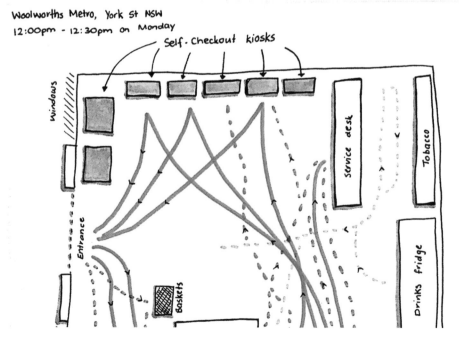

158 Case Study - Supermarket of the Future

This response to the Supermarket of the Future brief involves a competitor analysis and a spatial analysis of existing supermarkets. The envisioned solution is presented through mockups and visualisations.

Ideation Sketching

Design. Think. Make. Break. Repeat. 159

Final Visualisation

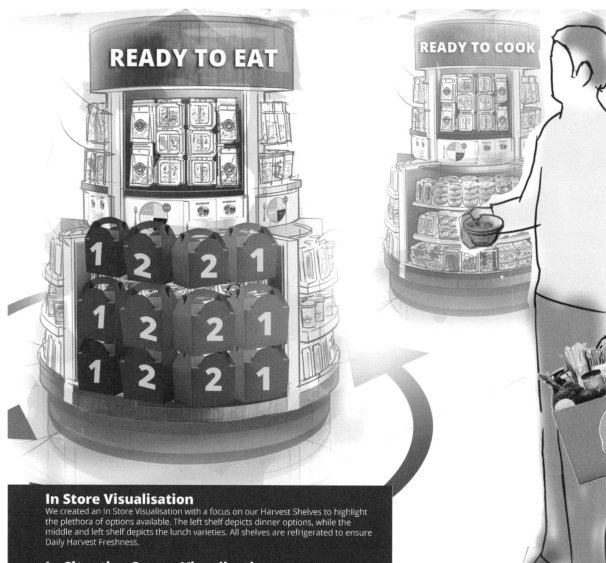

In Store Visualisation
We created an In Store Visualisation with a focus on our Harvest Shelves to highlight the plethora of options available. The left shelf depicts dinner options, while the middle and left shelf depicts the lunch varieties. All shelves are refrigerated to ensure Daily Harvest Freshness.

In-Situation Screen Visualisation
An initial In-Situation Screen Visualisation was wireframed and then designed into high fidelity to portray a user checking their Harvest selection while out and about. This was then developed for further user testing.

Architectural Floor Plan
We crafted an architectural floor plan to depict a potential layout that would optimise the front of house at a generic Woolworths Metro store. The primary focus was on the interactions for Harvest meals and checkout, while ensuring Harvest and regular shoppers had effortless access to the rest of the store as well as the Service Desk.

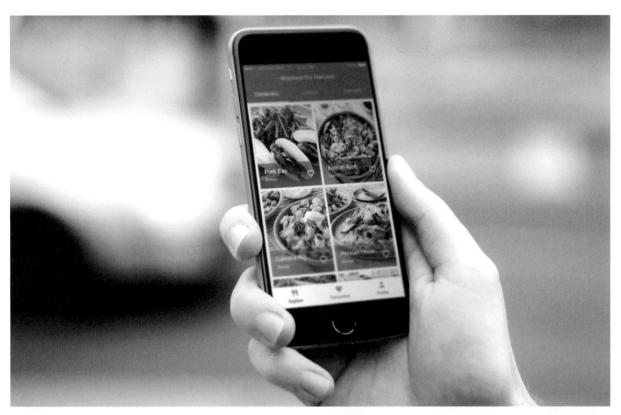

Design. Think. Make. Break. Repeat.

Templates

5 Whys

Notes sheet

Problem:

Why ...

Why ...

Why ...

Why ...

Why ...

Brainwriting 6-3-5

Notes sheet

Question/problem:

	Round 1	Round 2	Round 3	Round 4	Round 5	Round 6
Idea 1						
Idea 2						
Idea 3						

Design. Think. Make. Break. Repeat.

Business Model Canvas — Notes sheet

Key Partners	Key Activities		Value Propositions	Customer Relationships		Customer Segments
	Key Resources			Channels		
						Revenue Streams
Cost Structure						

Based on the original Business Model Canvas by Strategyzer AG: strategyzer.com

Business Model Experimentation

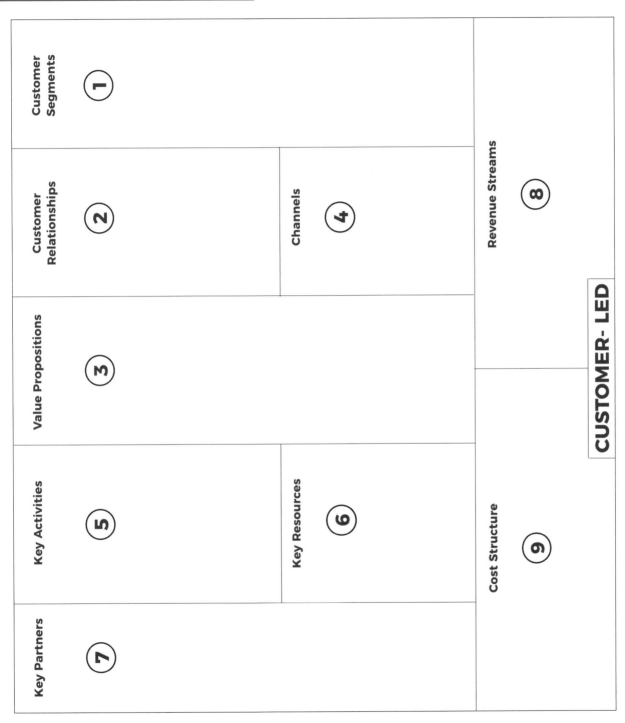

Based on the original Business Model Canvas by Strategyzer AG: strategyzer.com

Business Model Experimentation

6 Customer Segments		
7 Customer Relationships	5 Channels	9 Revenue Streams
8 Value Propositions		
2 Key Activities	3 Key Resources	**COST-DRIVEN**
4 Key Partners		1 Cost Structure

Based on the original Business Model Canvas by Strategyzer AG: strategyzer.com

Card Sorting

Data sheet

Project name:

Facilitator: Scribe:

Date: Time:

No. of participants: No. of cards:

Findings:

Observations	Participant comments	Final card groups

Recommendations for design:

Channel Mapping

Notes sheet

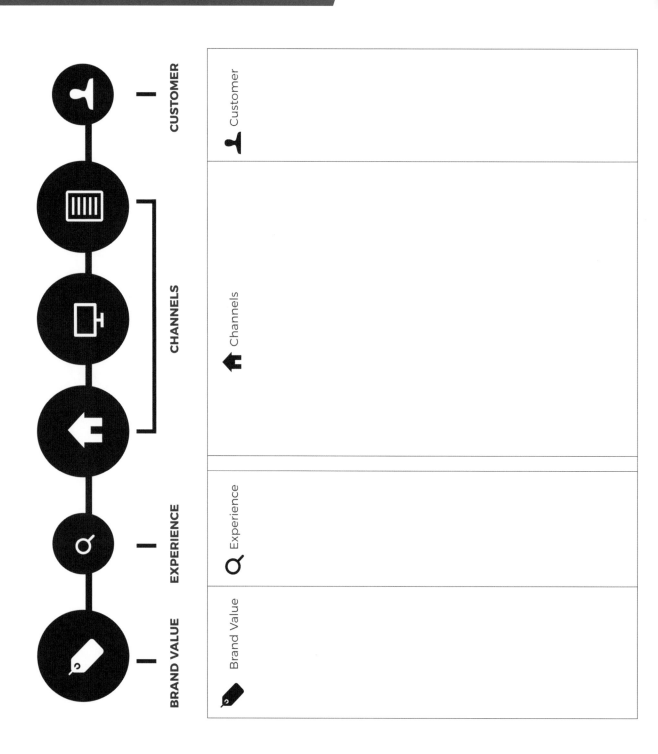

170 Templates

Competitor Analysis

Data sheet

Competitor	Competitor 1	Competitor 2	Competitor 3	Competitor 4
Customer segments				
Primary revenue stream				
Functionality offered				
Channels				
Key activities				
...				
...				
...				

Design. Think. Make. Break. Repeat.

Contextual Observation — Data sheet

Physical Behaviour (Facial expressions, gaze, gestures, posture, body language, vocal utterances, indicators of emotional state)					
Interface Part / Location					
User Goal / Task					

Decision Matrices
Structure guide

Criteria	Datum	Concept 1	Concept 2	Concept 3	...	...
e.g. fit with design brief						
...						
...						
...						
...						
...						
...						
...						
...						
...						
...						
...						
Number of pluses						
Number of minuses						
Overall total						

Design. Think. Make. Break. Repeat.

Design by Metaphor

Notes sheet

Explore the experience you are designing through your chosen metaphors. An example is included to get you started: redesigning a food delivery service through two different metaphors.

	Metaphor 1 **E.g. a swarm of bees**	**Metaphor 2** **E.g. swimming relay team**
Tell the metaphor's story	For example, if a food delivery service is like a swarm of bees, a team of worker bees simultaneously pick up orders and deliver to homes ...	
Elaborate the triggering concept		For example, the 'handover of the baton' could suggest an exchange between the person that delivers the food and the person ordering the food. What if someone drops the baton... ?
Look for new meanings for the concept		For example, the 'handover of the baton' concept could be interpreted as a symbolic gesture rather than a physical exchange.
Elaborate assumptions	For example, the swarm of bees metaphor highlights cooperative, parallel approaches to working together to create something.	For example, the swimming relay team metaphor highlights linear, sequential approaches to achieving a time-based goal.
Identify the unused part of the metaphor	For example, what happens when the Queen bee dies... ?	

Extreme Characters

Structure guide

Character Profile

Circle your extreme user type:

Good Samaritan / money-spinner / romantic / Trekkie

_____ is a _____ year old _____
<Name of character>   <Age>   <Man/woman/boy/girl>

who lives in _____ and _____
                    <Location>   <Works in/studies/does>

He/She _____
              <Values/interests/needs/motivations>

_____

_____

_____

_____

_____

Design. Think. Make. Break. Repeat.

Focus Groups

Data sheet

Project name:

Facilitator:　　　　　　　　　Scribe:　　　　　　　　　Date:　　　Time:

Draw your focus group here:

KEY　Table　[ ]

　　　　Participant　(P1)

　　　　Facilitator/ Scribe　(F) (S)

Participant comments

	Moderator question or topic	Participant responses
	Introduction and warm-up	
Focus 1	Draft questions and prompts	
Focus 2	Draft questions and prompts	
Focus 3	Draft questions and prompts	
	Thanks and conclusions	

Heuristic Evaluation — Notes sheet

Heuristic	Is the heuristic violated? How?	Severity
1. Visibility of system status The system should always keep users informed about what is going on, through appropriate feedback within reasonable time.		
2. Match between system and the real world The system should speak the users' language, with words, phrases and concepts familiar to the user, rather than system-oriented terms. Follow real-world conventions, making information appear in a natural and logical order.		
3. User control and freedom Users often choose system functions by mistake and will need a clearly marked 'emergency exit' to leave the unwanted state without having to go through an extended dialogue. Support undo and redo.		
4. Consistency and standards Users should not have to wonder whether different words, situations, or actions mean the same thing. Follow platform conventions.		
5. Error prevention Even better than good error messages is a careful design which prevents a problem from occurring in the first place.		
6. Recognition rather than recall Make objects, actions, and options visible. The user should not have to remember information from one part of the dialogue to another. Instructions for use of the system should be visible or easily retrievable whenever appropriate.		
7. Flexibility and efficiency of use Accelerators – unseen by the novice user – may often speed up the interaction for the expert user such that the system can cater to both inexperienced and experienced users. Allow users to tailor frequent actions.		
8. Aesthetic and minimalist design Dialogues should not contain information which is irrelevant or rarely needed. Every extra unit of information in a dialogue competes with the relevant units of information and diminishes their relative visibility.		
9. Help users recognise, diagnose and recover from errors Error messages should be expressed in plain language (no codes), precisely indicate the problem, and constructively suggest a solution.		
10. Help and documentation Even though it is better if the system can be used without documentation, it may be necessary to provide help and documentation. Any such information should be easy to search, focused on the user's task, list concrete steps to be carried out, and not be too large.		

Based on the 10 Usability Heuristics by Jakob Nielsen useit.com/papers/heuristic/heuristic_list.html

Online Ethnography — Data sheet

Source *(e.g. community or social media platform)* _____

Data record (E.g. quote)	Recorded observations (E.g. demographics, communication style, interests, social media habits, content published/shared)	Interpretation (E.g. what could this data mean)	Themes (E.g. what common patterns do you see)

Perceptual Maps — *Data sheet*

This questionnaire template is used to assess the consumer's perception on a given set of brands or products. For example, if the aim is to create a new breakfast cereal, the list should include well-known cereal products, such as Kellogg's All-Bran, Kellogg's Corn Flakes, Multi Grain Cheerios, Nestlé Chocapic, etc. Fill in the brand or product names below – one for each table. Identify pairs of opposing attributes, such as cheap versus expensive, and add them to each table.

Semantic differential scale

Ask your participants to rate each product according to their perception on the tables below.

1_____

	-5	-4	-3	-2	-1	0	1	2	3	4	5	
Cheap												Expensive

2_____

	-5	-4	-3	-2	-1	0	1	2	3	4	5	
Cheap												Expensive

Design. Think. Make. Break. Repeat.

Personas

Data sheet

Look for important variables in the data, and add labels for those variables at each end of the scale.
Map each individual from your research data onto the scale, using a letter to represent them.
Look for patterns: do the same letters appear side-by-side across multiple variables?

|—————————————————————————————|
Male   Female

|—————————————————————————————|
Younger   Older

|—————————————————————————————|

|—————————————————————————————|

|—————————————————————————————|

|—————————————————————————————|

|—————————————————————————————|

|—————————————————————————————|

Personas
Structure guide

Profile image: Sketch or photo

Persona type:

Name:

Occupation:

Age: **Gender:**

Backstory: Brief description of life story

Motivations: Why does the persona need to use the product/service?

Frustrations: What makes the persona feel frustrated or annoyed about the product/service?

Ideal experience / goals / aspirations / feelings:

Quote: Sum up the persona's experience

Reframing

Structure guide

A short description of your product or service:

_____

_____

Description with keywords changed:

Version 1 (change the person/people who are involved)

_____

_____

_____

Version 2 (change the setting where it happens)

_____

_____

_____

Version 3 (change the goal)

_____

_____

_____

Underlying goals

_____

_____

Rewritten problem statement

_____

_____

_____

Role-playing

Cards template

E.g. you are a traveller on holiday who has come to the ATM, to withdraw a large amount of cash in the local currency. You don't speak very good English.

You are a _____ (Person)

who has come to the _____ (Setting)

in order to: _____ (Goal)

(Backstory, details) _____

You are a _____ (Person)

who has come to the _____ (Setting)

in order to: _____ (Goal)

(Backstory, details) _____

You are a _____ (Person)

who has come to the _____ (Setting)

in order to: _____ (Goal)

(Backstory, details) _____

Design. Think. Make. Break. Repeat.

Scenarios

Structure guide

Story structure

Set the scene

Introduce the character

Problem/issue/need/motivation

Discovery/resolution

Narrative

Short title:

Key qualities
Make these visible in your concept

1

2

3

Science Fiction Prototyping

Steps	Narrative outline
1: Pick your science and build your world	
2: The scientific inflection point	
3: Ramifications of the science on people	
4: The human inflection point	
5: What did we learn?	

Based on: Johnson, B. D. (2011). Science fiction prototyping: Designing the future with science fiction. Synthesis Lectures on Computer Science, 3(1), 1-190.

Sketching

Sketch sheet

Take ten seconds for each of the steps in the table.

1 Sketch a building of any kind. The sketch should be very quick and low-fidelity.	2 Sketch a door.	3 Sketch a door that will let me and my cat in.
4 Sketch a door that will allow me to see who is on the other side before opening the door.	5 Sketch a door that will allow me to see who is on the other side but also allows me some privacy.	6 Sketch a door that will allow me to get through with a wheelchair.
7 Sketch a door that will allow me to bring large objects through it.	8 Sketch a door that allows for ventilation.	9 Sketch a door that allows me access to the roof.
10 Sketch a door that will inspire and amaze.	11 Sketch a door that will help me separate out those who walk through it.	12 Sketch a door to my backyard.

Reflect on how many doors you had to draw to satisfy each individual scenario. Are there any doors that satisfy multiple of the scenarios?

Review how many doors fit into the building from step one?

Source: UX Australia 2014 talk by Supriya Perera

Storyboards

Sketch sheet

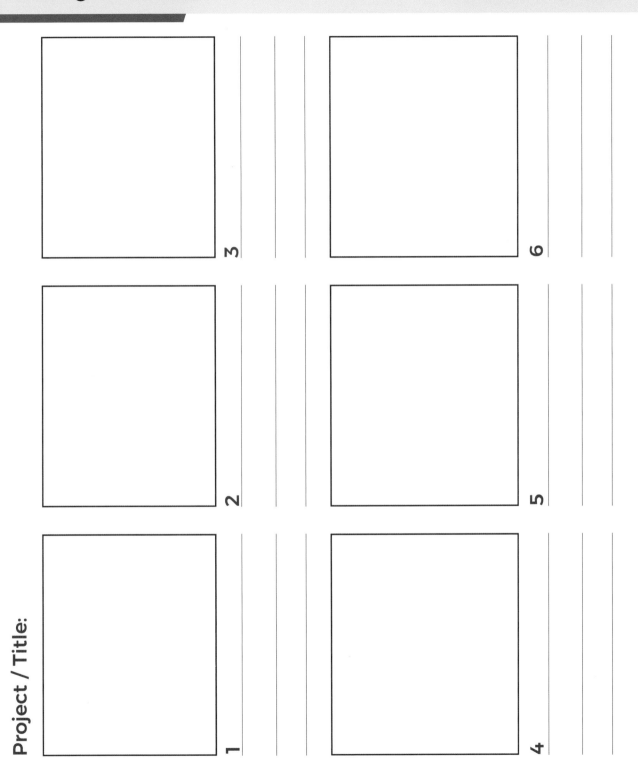

Design. Think. Make. Break. Repeat.

Thematic Analysis — Data sheet

Total number of interviews:

Theme	Example quotes or excerpts	No. of people to express theme e.g. 3 interviewees	No. of references e.g. 73 references	Notes/Comments What does this mean for your design?

Value Proposition Canvas — *Notes sheet*

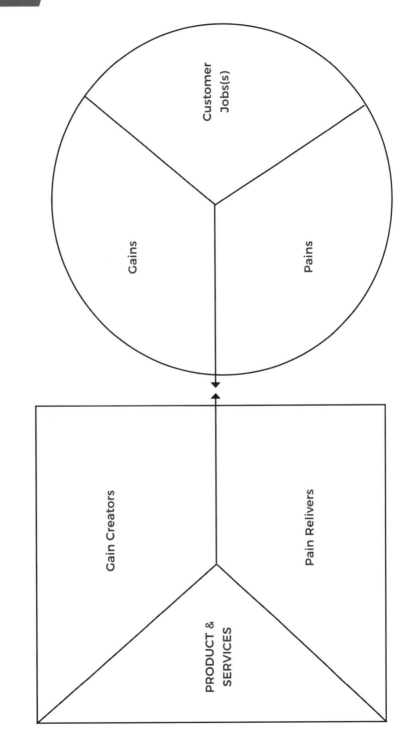

Based on the original Value Proposition Canvas by Strategyzer AG: strategyzer.com

Think-aloud Protocol

	Verbal Protocol	Interface Part / Location	User Goal / Task

Usability Testing
Data sheet

Notetaker _____

Participant # _____

Tested product (e.g. website URL): _____

Task(s) (Enter a brief description for each task)	Success 0 = Not completed 1 = Completed with difficulty or help 2 = Easily completed	Time to complete	Number of Errors	Notes/Observations (Note why the user was successful or not successful, e.g., wrong pathways, confusing page layout, navigation issues, terminology)
#1:				
#2:				
#3:				
#4:				
#5:				

Usability Testing — Data sheet

System Usability Scale

Participant #_____

		strongly disagree 1	2	3	4	strongly agree 5
1	I think that I would like to use this system frequently.					
2	I found the system unnecessarily complex.					
3	I thought the system was easy to use.					
4	I think that I would need the support of a technical person to be able to use this system.					
5	I found the various functions in this system were well-integrated.					
6	I thought there was too much inconsistency in this system.					
7	I would imagine that most people would learn to use this system very quickly.					
8	I found the system very cumbersome to use.					
9	I felt very confident using the system.					
10	I needed to learn a lot of things before I could get going with this system.					

Source: Brooke, J. (1996). SUS-A quick and dirty usability scale. Usability evaluation in industry, 189(194), 4-7.

Consent Form

I agree to participate in the study conducted and recorded by _____.

I agree to:
☐ The session being audio/video-recorded (cross out as appropriate)
☐ The use of photographs and video recordings for the purpose of documenting the findings from this study

I understand that the information collected in this study is for research purposes only and that my name and image will not be used for any other purpose. I relinquish any rights to the recording.

I understand that participation in this usability study is voluntary and I agree to immediately raise any concerns or areas of discomfort during the session with the study administrator.

I confirm that I have read and understand the information on this form and that any questions I might have about the session have been answered.

Date:_____

Please print your name: _____

Please sign your name:_____

Thank you! We appreciate your participation.

User Journey Mapping — *Structure guide*

Stages	Activities	Thoughts & Emotions	Touch-points	+	Pains & Gains Map	−

User Profiles

Structure guide

Fill in the identified user profile in the top row, e.g. 'Budget traveler', 'Frequent flyer', etc.
Add additional attributes in the first column, e.g. skills, technology expertise, attitudes, roles, responsibilities, etc.

User Profile Type	Age range	Motivation	Needs				

194 Templates

Wireframing

Sketch sheet

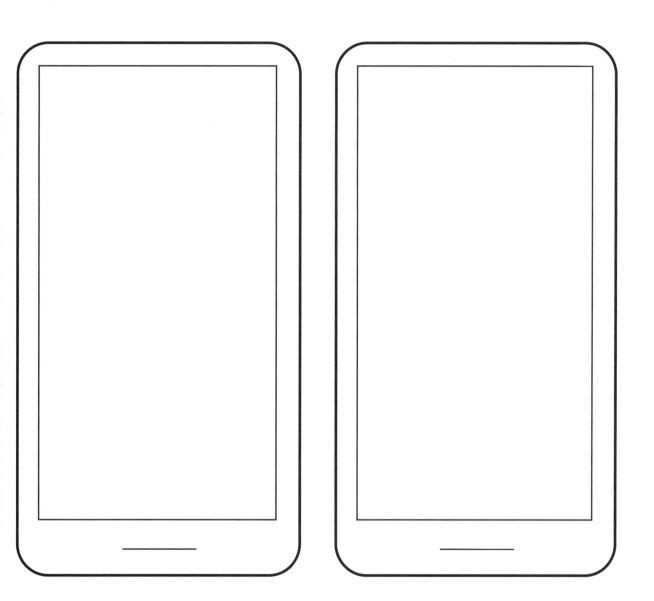

Design. Think. Make. Break. Repeat.

Design Team

Martin Tomitsch

Dr Martin Tomitsch is Associate Professor and Chair of Design at the University of Sydney School of Architecture, Design and Planning and Director of the Design Lab. He received his Ph.D. in informatics from the Vienna University of Technology. His research focuses on the role of design in shaping the interactions between people and technology. He has been teaching interaction design in university programs since 2004. He is state co-chair of the Australian Computer-Human Interaction Special Interest Group (CHISIG), visiting lecturer at the Vienna University of Technology's Research Group for Industrial Software (INSO), and visiting professor at the Central Academy of Fine Arts, Beijing.

Cara Wrigley

Dr Cara Wrigley is Associate Professor of Design Innovation at the University of Sydney, residing in the Design Lab - an interdisciplinary research group within the School of Architecture, Design and Planning. She is an industrial designer who is actively researching the value that design holds in business – specifically through the creation of strategies to design business models which lead to emotive customer engagement. Her primary research interest is in the application and adoption of design innovation methods by various industry sectors to better address latent customer needs. Her work has crossed research boundaries and appears in a wide range of disciplinary publications.

Madeleine Borthwick

Madeleine Borthwick is an Associate Lecturer at the University of Sydney's School of Architecture, Design and Planning. She is responsible for teaching and coordinating design subjects that include 'User Experience Design', 'Interaction Design', 'Design Processes & Methods', and '3D Modelling and Fabrication'. Additionally, Madeleine is a practicing interaction designer and the director of Kiss the Frog Australia: a consultancy specialising in the design of creative multimedia-based experiences for museums and visitor centres. Her background is in interaction design (Delft University of Technology, MSc) and industrial design (University of Technology Sydney, Hons).

Naseem Ahmadpour

Dr Naseem Ahmadpour is a Lecturer in design computing at The University of Sydney. She received her Ph.D. in design and interaction from Polytechnique Montréal. Naseem's research is interdisciplinary and broadly focused on design for wellbeing. Specifically, she investigates new design possibilities that meet basic human needs and values and therefore enhance motivation and the ability to self-regulate; particularly in the area of health. Naseem has held positions at the Swinburne University of Technology and industries including Bombardier Aerospace and has been a visiting scholar at the Delft University of Technology. She has published in leading design and human factors journals and conferences including Applied Ergonomics, Ergonomics, and Design Research Society.

Jessica Frawley

Dr Jessica Frawley is a Lecturer in educational innovation and an Honorary Associate of the School of Architecture, Design and Planning at the University of Sydney. Having a cross-disciplinary background, with degrees in both information technology and the humanities, Jessica's work focuses on designing and understanding technologies from human perspectives. She has worked as a researcher and designer in a range of academic, commercial, and government settings, but is especially focused on education and technologies for learning. Her research and teaching have been recognised through multiple awards and are featured in a wide range of publications.

A. Baki Kocaballi

Dr A. Baki Kocaballi is a Postdoctoral Research Fellow in artificial intelligence and interaction design at Macquarie University. He completed his Ph.D. in interaction design at the University of Sydney in 2013. Before undertaking his doctoral study, he completed a Master's degree in information systems at the Middle East Technical University. He has extensive knowledge and experience in designing and developing educational and medical applications. He has received several academic and artistic awards and recognitions. His research interests include situated and relational design approaches, actor-network theory, artificial intelligence, user experience, e-health, conversational interfaces, ideation, and participatory design.

Claudia Núñez-Pacheco

Claudia Núñez-Pacheco is a design researcher and Ph.D. candidate in the Design Lab at the University of Sydney. Her research investigates how bodily ways of knowing can be used as crafting materials for design ideation, evaluation, insight and empathy. In her research journey, Claudia has engaged in a multidisciplinary exploration that merges human-computer interaction (HCI) and design methods with tools from experiential psychology. Claudia has been awarded twice by the National Commission for Scientific and Technological Research Scholarship (Chile), in addition to disseminating her research through various international HCI and design publications.

Karla Straker

Dr Karla Straker is an Early Career Development Fellow, in the Design Lab, located in the School of Architecture, Design and Planning, at the University of Sydney. She has a Bachelor of Design (industrial design) and a Ph.D. from the Queensland University of Technology. Her research is in a cross-disciplinary setting and explores the design of digital channel engagements, through theoretical approaches from the fields of design, psychology, marketing and information systems. Her research aims to understand how relationships with customers can be built and sustained through a deeper understanding of emotions. In her research work, she emphasises the design and evaluation of new approaches to the field of design innovation. Karla has published in a variety of journals spanning the fields of marketing, interactive technology, business, and design.

Lian Loke

Dr Lian Loke is a Senior Lecturer in design and computation at the University of Sydney, and Director of the Master of Interaction Design and Electronic Arts. She received her Ph.D. in interaction design from the University of Technology Sydney. Her research explores the aesthetics of interaction and creative approaches to designing for the user experience of interactive technologies informed by dance, performance, and somatics. She has published in leading design and human-computer interaction journals, for example, Transactions in Computer-Human Interaction, and International Journal of Design.

Other Contributors

George Peppou
George Peppou is a Ph.D. candidate in the Design Lab at the University of Sydney studying the role of design thinking in the commercialisation of complex technologies. George works closely with industry mentoring early-stage technology startups.

Dagmar Reinhardt
Dr Dagmar Reinhardt is the Program Director of the Bachelor of Architecture and Environments at the School of Architecture, Design and Planning, the University of Sydney. Her research extends architectural performance and 'performativity' towards design research and cross-disciplinary practice with audio-acoustics, and structural engineering.

Alex Garrett
Dr Alex Garrett is a Ph.D. Candidate at the University of Sydney. He earned his Bachelor (First Class Honours) in industrial design in 2013. Alex is currently undertaking his Ph.D. in design-led innovation, a journey due for completion in 2018. Alex has published in several leading design, business and innovation conferences, and journals.

Glossary

Affordance:
Affordance is a relational concept explaining how a particular action potential emerges out of a match between the capabilities of a subject and the properties of an object. For example, if an object has a sufficiently large flat surface at a suitable height then it may afford sit-on-ability.

Concept mapping:
Concept maps are used for visually representing, organising and structuring ideas on a topic. Individual concepts are placed in circles and connected with other relevant concepts by lines or arrows with brief descriptive labels describing the relation. One common organisation method is to order the concepts in a hierarchical order.

Conceptual model:
A conceptual model is the overall image representing the working mechanism of a product. It is constructed using a product's features and accompanying documentation. A good conceptual model should match with the mental models of users in order to facilitate an error-free user experience.

Convergent thinking:
Convergent thinking aims to narrow down the solution space by eliminating less feasible ideas. It typically takes place after a process of divergent thinking producing many alternative ideas. For example, analysing the user testing of different prototypes involve convergent thinking process, resulting in reduced number of alternatives.

Divergent thinking:
Divergent thinking aims to expand the solution space by generating as many ideas as possible. It is followed by a process of convergent thinking to reduce the number of alternative ideas. The focus is usually upon the quantity of ideas rather than quality. Brainstorming is the most widely used divergent thinking method.

Gestalt:
Gestalt or holism refers to the idea that several seemingly separate items can be perceived as a whole single entity or system if they are organised according to any one of six *Gestalt* principles: similarity, continuation, closure, proximity, figure/ground and symmetry and order.

Interface:
From a high-level perspective, an interface is a system that connects the analogue with the digital. User interfaces (UI) can be defined as the interactive sections of technology artefacts that mediate the communication between hardware and people. A good UI enables natural and transparent interactions, allowing users to focus on their everyday tasks (such as writing and sending an email), without having to worry about the technical aspects behind this action.

Likert scales:
A Likert scale is a method for measuring a person's response to a statement. Various scales exist, however mostccommonly an even-point scale is used to question the extent to which the person agrees or disagrees with the statement. This usually requires a person to choose from five to seven pre-coded options with neutral being the middle point.

Mental model:
Mental models are people's generalisations about how things around them function. In design, they allow a user to understand and interact with a new product based on her previous experience with similar products. Mental models can be incomplete and vague, causing errors. Therefore, designers aim to reduce any mismatches between users' mental models and the features of their product.

Participatory design:
Participatory design emerged out of a concern that people whose lives are going to be affected by the introduction of a new technology need to have a say about the design of those technologies. Participatory design empowers the stakeholders of a design project by offering various methods that allow the alternative ideas to be developed, prototyped and tested collectively.

Patterns:
A pattern is a generally repeatable solution to a commonly occurring problem within a given context. It is not a finished design that can be directly applied; rather patterns also require interpretation to determine how to apply them properly. Patterns occur in many different design domains, including interface design, architecture and software development.

Semantic differential scales:
Semantic differential scales enable an assessment of non-material and subjective traits of an experience by using and contrasting semantic terms, such as pleasant versus unpleasant. These traits are represented on the opposite ends of a continuum, allowing the individual to make a connection between their experience and those traits somewhere along the scale.

Stakeholders:
A stakeholder is a person, group, or organisation that has a vested interest in or concern with the design problem or solution.

Tacit knowledge:
Tacit knowledge is accumulated through experience and underpins much of what we do or say. It is more than we can typically verbalise or write down. Embodied methods such as role-playing or bodystorming rely on tacit knowledge of how to act, perceive and feel in the world.

Credits

Project lead:
Martin Tomitsch

Project coordination:
Madeleine Borthwick

Introduction:
Martin Tomitsch and Cara Wrigley

Methods and exercises:
Martin Tomitsch, Cara Wrigley, Madeleine Borthwick, Naseem Ahmadpour, Jessica Frawley, A. Baki Kocaballi, Claudia Núñez-Pacheco, Karla Straker, Lian Loke, with contributions from Alex Garrett, George Peppou, Dagmar Reinhardt

Editing of methods and exercises:
Martin Tomitsch and Madeleine Borthwick
Verity Borthwick (1st reprint)

Templates:
If not otherwise stated, Martin Tomitsch, Cara Wrigley, Madeleine Borthwick, Naseem Ahmadpour, Jessica Frawley, A. Baki Kocaballi, Claudia Núñez-Pacheco, Karla Straker, Lian Loke

Layout and graphic design:
Matthew Fehlberg, Soomin Lee (1st reprint)

Sketches:
Nikash Singh

Diagrams:
A. Baki Kocaballi

Cover illustration:
Nikash Singh

Production:
BIS Publishers

This book is based on the Sydney Design Thinking Toolkit project, which was a 2016 Educational Innovation initiative funded by the University of Sydney's Strategic Education Grants scheme.

www.designthinkmakebreakrepeat.com
contact@designthinkmakebreakrepeat.com

---

5 Whys, p.18: Art Poskanzer, CC BY 2.0, https://www.flickr.com/photos/posk/8333973575/
A/B Testing, p.20: Robert Couse-Baker, CC BY 2.0, www.flickr.com/photos/29233640@N07/12596035923/
Affinity Diagramming, p.22: Chris Green, Rachel Montgomery, Natalia Gulbranson-Diaz
Autobiographical Diaries, p.24: Cecile Tran
Bodystorming, p.26: Stephen P. Carmody, CC BY 2.0, https://www.flickr.com/photos/scarms/34633589670/
Brainwriting 6-3-5, p.28: kellywritershouse, CC BY 2.0, https://www.flickr.com/photos/kellywritershouse/5529146953/
Business Model Canvas, p.30: http://firstround.com/review/To-Go-Lean-Master-the-Business-Model-Canvas/
Business Model Experimentation, p.32: Alexander Osterwalder, CC BY 2.0, https://www.flickr.com/photos/osterwalder/4184663774/
Card Sorting, p.34: Sarah Brooks, CC BY 2.0, https://www.flickr.com/photos/foodclothingshelter/3511435109/
Cartographic Mapping, p.36: A. Baki Kocaballi
Channel Mapping, p.38: Gauthier Delecroix, CC BY 2.0, https://www.flickr.com/photos/gauthierdelecroix/33425765802/
Co-design Workshops, p.40: Eleonora Mencarini - meSch project, CC BY 2.0, https://www.flickr.com/photos/meschproject/10398447836/
Competitor Analysis, p.42: Ashlee Martin, CC BY 2.0, https://www.flickr.com/photos/ashl33/3391702714/
Contextual Observation, p.44: U.S. Fish and Wildlife Service Southeast Region, CC BY 2.0, https://www.flickr.com/photos/usfwssoutheast/8077209942/
Cultural Probes, p.46: Gunnar Bothner-By, CC BY 2.0, https://www.flickr.com/photos/gcbb/3234180323/ https://www.flickr.com/photos/gcbb/3235030262/
Decision Matrices, p.48: Siaron James, CC BY 2.0, https://www.flickr.com/photos/59489479@N08/30040305296/
Design by Metaphor, p.50: U.S. Department of Agriculture, CC BY 2.0, https://www.flickr.com/photos/usdagov/27264129734/
Design Critique, p.52: Drew, CC BY 2.0, https://www.flickr.com/photos/drew-harry/5392135365/
Direct Experience Storyboards, p.54: Matt Fehlberg, Francesca Serpollini, Mackenzie Etherington, Alex Elton-Pym
Empathic Modelling, p.56: Danijel Šivinjski, Public Domain Dedication (CC0), https://www.flickr.com/photos/blindfields/34132870980/
Experience Prototyping, p.58: amy gizienski, CC BY 2.0, https://www.flickr.com/photos/agizienski/3778965891/
Experience Sampling, p.60: Hans Luthart, CC BY 2.0, https://www.

Image Sources and References

flickr.com/photos/vrijstaat/25163862074/

Extreme Characters, p.62: Dawei Zhou

Focus Groups, p.64: Universidad Magallanes, Public Domain Mark 1.0, https://www.flickr.com/photos/umag/33152726634/

Forced Associations, p.66: Lottie, Public Domain Dedication (CC0), https://www.flickr.com/photos/milkyfactory/16979768317/

Future Workshops, p.68: Mei Anne Mendoza, CC BY 2.0, https://www.flickr.com/photos/ham_phtgrphy/3774337607/

Group Passing, p.70: Drew, CC BY 2.0, https://www.flickr.com/photos/drew-harry/5392729156/

Hero Stories, p.72: Mark Gunn, CC BY 2.0, https://www.flickr.com/photos/mark-gunn/14601979455/

Heuristic Evaluation, p.74: Image catalog, Public Domain Dedication (CC0), https://www.flickr.com/photos/image-catalog/18692128651/

Interaction Relabelling, p.76: Ian D. Keating, CC BY 2.0, https://www.flickr.com/photos/ian-arlett/24171851760/

Interviews, p.78: Alper Çuğun, CC BY 2.0, https://www.flickr.com/photos/alper/9516543726/

KJ Brainstorming, p.80: Katarzyna Stawarz, CC BY 2.0, https://www.flickr.com/photos/flk/5481461561/

Laddering, p.82: Hamza Butt, CC BY 2.0, https://www.flickr.com/photos/149902454@N08/35417844812/

Low-fidelity Prototyping, p.84: Lindy Zubairy, CC BY-ND 2.0, https://www.flickr.com/photos/rankinmiss/8022928786/

Mapping Space, p.86: Lauren, CC BY 2.0, https://www.flickr.com/photos/seelauren/2826595563/

Mindmapping (WWWWWH), p.88: Tom Henderson, CC BY 2.0, https://www.flickr.com/photos/infrarad/224075209/

Mock-ups, p.90: WorldSkills UK, CC BY 2.0, https://www.flickr.com/photos/worldskillsteamuk/8189707845/

Mood Boards, p.92: Jordanhill School D&T Dept, CC BY 2.0, https://www.flickr.com/photos/designandtechnologydepartment/3973280088/

Online Ethnography, p.94: alaina buzas, CC BY 2.0, https://www.flickr.com/photos/alaina_marie/4186159031/

Perceptual Maps, p.96: Rasheeda Ahmed – Jordanhill School D&T Dept, CC BY 2.0, https://www.flickr.com/photos/designandtechnologydepartment/4032186959/

Persona-based Walkthroughs, p.98: Mish Sukharev, CC BY 2.0, https://www.flickr.com/photos/mishism/3877391184/

Personas, p.100: Matthew Fehlberg

Questionnaires, p.102: Christine und Hagen Graf, CC BY 2.0, https://www.flickr.com/photos/hagengraf/15667793640/

Reframing, p.104: Katarína Chovancová, CC BY-SA 2.0, https://www.flickr.com/photos/128196805@N03/16798815406/

Research Visualisation, p.106: Matthew Fehlberg

Role-playing, p.108: Emily Tulloh, CC BY 2.0, https://www.flickr.com/photos/131402175@N07/16371953780/

Scenarios, p.110: Garrett, CC BY 2.0, https://www.flickr.com/photos/garrett-ewp/5811716939/

Science Fiction Prototyping, p.112: Canadian Film Centre, CC BY 2.0, https://www.flickr.com/photos/cfccreates/10578806904/

Service Blueprints, p.114: pin add, CC BY 2.0, https://www.flickr.com/photos/pinadd/2858659917/

Sketching, p.116: Peter Lindberg, CC BY 2.0, https://www.flickr.com/photos/plindberg/2557393147/

Sketchnoting, p.118: Soomin Lee

Storyboarding, p.120: Chinmay Kulkarni

Thematic Analysis, p.122: Vijay Chennupati, CC BY 2.0, https://www.flickr.com/photos/vijay_chennupati/34255520505/

Think-aloud Protocol, p.124: City University Interaction Lab, CC BY 2.0, https://www.flickr.com/photos/cinteractionlab/4557093503/

Usability Testing, p.126: City University Interaction Lab, CC BY 2.0, https://www.flickr.com/photos/cinteractionlab/4557114137/

User Journey Mapping, p.128: Rawpixel.com, Public Domain Dedication (CC0), https://www.flickr.com/photos/byrawpixel/32881021611/

User Profiles, p.130: Kate Bookallil

Value Proposition Canvas, p.132: Kevin Dooley, CC BY 2.0, https://www.flickr.com/photos/pagedooley/6575053747/

Video Prototyping, p.134: Adrian Clark, CC BY 2.0, https://www.flickr.com/photos/adriandc/2185215183/

Wireframing, p.136: Chinmay Kulkarni

Autonomous Vehicles, p.140: Martin Tomitsch

Designing Space Travel, p.141: NASA, https://unsplash.com/photos/yZygONrUBe8

Museum Visitor Experience, p.142: Martin Tomitsch

Supermarket of the Future, p.143: Dean Hochman, CC BY 2.0, https://www.flickr.com/photos/deanhochman/14961829054/

Think-aloud Protocol and Usability Testing templates, p.190–191: adapted from www.usability.gov

Pencil icon by Erin Standley, CC BY 3.0 US, https://thenounproject.com/search/?q=post+it&i=17650

Light bulb icon by Consumer Financial Protection Bureau, CC0 1.0, https://thenounproject.com/search/?q=bulb&i=89676

Wrench icon by Erin Standley, CC BY 3.0 US, https://thenounproject.com/term/wrench/17679/

Repeat and Break icons by Nikash Singh